KINETIC MODEL

TASK CARD SERIES

Conceived and written by
RON MARSON

Illustrated by
PEG MARSON

TOPS LEARNING SYSTEMS

10970 S Mulino Road
Canby OR 97013
Website: topscience.org
Fax: 1 (503) 266-5200

Oh, those pesky COPYRIGHT RESTRICTIONS !

Dear Educator,

TOPS is a nonprofit organization dedicated to educational ideals, not our bottom line. We have invested much time, energy, money, and love to bring you this excellent teaching resource.

And we have carefully designed this book to run on simple materials you already have or can easily purchase. If you consider the depth and quality of this curriculum amortized over years of teaching, it is dirt cheap, orders of magnitude less than prepackaged kits and textbooks.

Please honor our copyright restrictions. We are a very small company, and book sales are our life blood. When you buy this book and use it for your own teaching, you sustain our publishing effort. If you give or "loan" this book or copies of our lessons to other teachers, with no compensation to TOPS, you squeeze us financially, and may drive us out of business. Our well-being rests in your hands.

What if you are excited about the terrific ideas in this book, and want to share them with your colleagues? What if the teacher down the hall, or your homeschooling neighbor, is begging you for good science, quick! We have suggestions. Please see our *Purchase and Royalty Options* below.

We are grateful for the work you are doing to help shape tomorrow. We are honored that you are making TOPS a part of your teaching effort. Thank you for your good will and kind support.

Sincerely, *Ron Marson*

Purchase and Royalty Options:

Individual teachers, homeschoolers, libraries:

PURCHASE option: If your colleagues are asking to borrow your book, please ask them to read this copyright page, and to contact TOPS for our current catalog so they can purchase their own book. We also have an **online catalog** that you can access at www.topscience.org.

If you are reselling a **used book** to another classroom teacher or homeschooler, please be aware that this still affects us by eliminating a potential book sale. We do not push "newer and better" editions to encourage consumerism. So we ask seller or purchaser (or both!) to acknowledge the ongoing value of this book by sending a contribution to support our continued work. Let your conscience be your guide.

Honor System ROYALTIES: If you wish to make copies from a library, or pass on copies of just a few activities in this book, please calculate their value at 50 cents (25 cents for homeschoolers) per lesson per recipient. Send that amount, or ask the recipient to send that amount, to TOPS. We also gladly accept donations. We know life is busy, but please do follow through on your good intentions promptly. It will only take a few minutes, and you'll know you did the right thing!

Schools and Districts:

You may wish to use this curriculum in several classrooms, in one or more schools. Please observe the following:

PURCHASE option: Order this book in quantities equal to the number of target classrooms. If you order 5 books, for example, then you have unrestricted use of this curriculum in any 5 classrooms per year for the life of your institution. You may order at these quantity discounts:

2-9 copies: 90% of current catalog price + shipping.

10+ copies: 80% of current catalog price + shipping.

ROYALTY option: Purchase 1 book *plus* photocopy or printing rights in quantities equal to the number of designated classrooms. If you pay for 5 Class Licenses, for example, then you have purchased reproduction rights for any 5 classrooms per year for the life of your institution.

1-9 Class Licenses: 70% of current book price per classroom.

10+ Class Licenses: 60% of current book price per classroom.

Workshops and Training Programs:

We are grateful to all of you who spread the word about TOPS. Please limit duplication to only those lessons you will be using, and collect all copies afterward. No take-home copies, please. Copies of copies are prohibited. Ask us for a free shipment of as many current **TOPS Ideas** catalogs as you need to support your efforts. Every catalog contains numerous free sample teaching ideas.

ISBN 0-941008-84-3

CONTENTS

A TOPS Model for Effective Science Teaching...

If science were only a set of explanations and a collection of facts, you could teach it with blackboard and chalk. You could assign students to read chapters and answer the questions that followed. Good students would take notes, read the text, turn in assignments, then give you all this information back again on a final exam. Science is traditionally taught in this manner. Everybody learns the same body of information at the same time. Class togetherness is preserved.

But science is more than this.

Science is also process — a dynamic interaction of rational inquiry and creative play. Scientists probe, poke, handle, observe, question, think up theories, test ideas, jump to conclusions, make mistakes, revise, synthesize, communicate, disagree and discover. Students can understand science as process only if they are free to think and act like scientists, in a classroom that recognizes and honors individual differences.

Science is *both* a traditional body of knowledge *and* an individualized process of creative inquiry. Science as process cannot ignore tradition. We stand on the shoulders of those who have gone before. If each generation reinvents the wheel, there is no time to discover the stars. Nor can traditional science continue to evolve and redefine itself without process. Science without this cutting edge of discovery is a static, dead thing.

Here is a teaching model that combines the best of both elements into one integrated whole. It is only a model. Like any scientific theory, it must give way over time to new and better ideas. We challenge you to incorporate this TOPS model into your own teaching practice. Change it and make it better so it works for you.

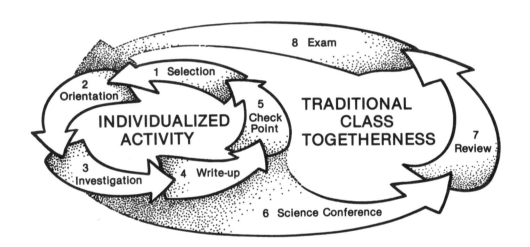

1. SELECTION

Doing TOPS is as easy as selecting the first task card and doing what it says, then the second, then the third, and so on. Working at their own pace, students fall into a natural routine that creates stability and order. They still have questions and problems, to be sure, but students know where they are and where they need to go.

Students generally select task cards in sequence because new concepts build on old ones in a specific order. There are, however, exceptions to this rule: students might *skip* a task that is not challenging; *repeat* a task with doubtful results; *add* a task of their own design to answer original "what would happen if" questions.

2. ORIENTATION

Many students will simply read a task card and immediately understand what to do. Others will require further verbal interpretation. Identify poor readers in your class. When they ask, "What does this mean?" they may be asking in reality, "Will you please read this card aloud?"

With such a diverse range of talent among students, how can you individualize activity and still hope to finish this module as a cohesive group? It's easy. By the time your most advanced students have completed all the task cards, including the enrichment series at the end, your slower students have at least completed the basic core curriculum. This core provides the common background so necessary for meaningful discussion, review and testing on a class basis.

3. INVESTIGATION

Students work through the task cards independently and cooperatively. They follow their own experimental strategies and help each other. You encourage this behavior by helping students only *after* they have tried to help themselves. As a resource person, you work to stay *out* of the center of attention, answering student questions rather than posing teacher questions.

When you need to speak to everyone at once, it is appropriate to interrupt individual task card activity and address the whole class, rather than repeat yourself over and over again. If you plan ahead, you'll find that most interruptions can fit into brief introductory remarks at the beginning of each new period.

4. WRITE-UP

Task cards ask students to explain the "how and why" of things. Write-ups are brief and to the point. Students may accelerate their pace through the task cards by writing these reports out of class.

Students may work alone or in cooperative lab groups. But each one must prepare an original write-up. These must be brought to the teacher for approval as soon as they are completed. Avoid dealing with too many write-ups near the end of the module, by enforcing this simple rule: each write-up must be approved *before* continuing on to the next task card.

5. CHECK POINT

The student and teacher evaluate each write-up together on a pass/no-pass basis. (Thus no time is wasted haggling over grades.) If the student has made reasonable effort consistent with individual ability, the write-up is checked off on a progress chart and included in the student's personal assignment folder or notebook kept on file in class.

Because the student is present when you evaluate, feedback is immediate and effective. A few seconds of this direct student-teacher interaction is surely more effective than 5 minutes worth of margin notes that students may or may not heed. Remember, you don't have to point out every error. Zero in on particulars. If reasonable effort has not been made, direct students to make specific improvements, and see you again for a follow-up check point.

A responsible lab assistant can double the amount of individual attention each student receives. If he or she is mature and respected by your students, have the assistant check the even-numbered write-ups while you check the odd ones. This will balance the work load and insure that all students receive equal treatment.

6. SCIENCE CONFERENCE

After individualized task card activity has ended, this is a time for students to come together, to discuss experimental results, to debate and draw conclusions. Slower students learn about the enrichment activities of faster students. Those who did original investigations, or made unusual discoveries, share this information with their peers, just like scientists at a real conference. This conference is open to films, newspaper articles and community speakers. It is a perfect time to consider the technological and social implications of the topic you are studying.

7. READ AND REVIEW

Does your school have an adopted science textbook? Do parts of your science syllabus still need to be covered? Now is the time to integrate other traditional science resources into your overall program. Your students already share a common background of hands-on lab work. With this shared base of experience, they can now read the text with greater understanding, think and problem-solve more successfully, communicate more effectively.

You might spend just a day on this step or an entire week. Finish with a review of key concepts in preparation for the final exam. Test questions in this module provide an excellent basis for discussion and study.

8. EXAM

Use any combination of the review/test questions, plus questions of your own, to determine how well students have mastered the concepts they've been learning. Those who finish your exam early might begin work on the first activity in the next new TOPS module.

Now that your class has completed a major TOPS learning cycle, it's time to start fresh with a brand new topic. Those who messed up and got behind don't need to stay there. Everyone begins the new topic on an equal footing. This frequent change of pace encourages your students to work hard, to enjoy what they learn, and thereby grow in scientific literacy.

GETTING READY

Here is a checklist of things to think about and preparations to make before your first lesson.

☐ Decide if this TOPS module is the best one to teach next.

TOPS modules are flexible. They can generally be scheduled in any order to meet your own class needs. Some lessons within certain modules, however, do require basic math skills or a knowledge of fundamental laboratory techniques. Review the task cards in this module now if you are not yet familiar with them. Decide whether you should teach any of these other TOPS modules first: *Measuring Length, Graphing, Metric Measure, Weighing* or *Electricity* (before *Magnetism*). It may be that your students already possess these requisite skills or that you can compensate with extra class discussion or special assistance.

☐ Number your task card masters in pencil.

The small number printed in the lower right corner of each task card shows its position within the overall series. If this ordering fits your schedule, copy each number into the blank parentheses directly above it at the top of the card. Be sure to use pencil rather than ink. You may decide to revise, upgrade or rearrange these task cards next time you teach this module. To do this, write your own better ideas on blank 4 x 6 index cards, and renumber them into the task card sequence wherever they fit best. In this manner, your curriculum will adapt and grow as you do.

☐ Copy your task card masters.

You have our permission to reproduce these task cards, for as long as you teach, with only 1 restriction: please limit the distribution of copies you make to the students you personally teach. Encourage other teachers who want to use this module to purchase their *own* copy. This supports TOPS financially, enabling us to continue publishing new TOPS modules for you. For a full list of task card options, please turn to the first task card masters numbered "cards 1-2."

☐ Collect needed materials.

Please see the opposite page.

☐ Organize a way to track completed assignment.

Keep write-ups on file in class. If you lack a vertical file, a box with a brick will serve. File folders or notebooks both make suitable assignment organizers. Students will feel a sense of accomplishment as they see their file folders grow heavy, or their notebooks fill up, with completed assignments. Easy reference and convenient review are assured, since all papers remain in one place.

Ask students to staple a sheet of numbered graph paper to the inside front cover of their file folder or notebook. Use this paper to track each student's progress through the module. Simply initial the corresponding task card number as students turn in each assignment.

☐ Review safety procedures.

Most TOPS experiments are safe even for small children. Certain lessons, however, require heat from a candle flame or Bunsen burner. Others require students to handle sharp objects like scissors, straight pins and razor blades. These task cards should not be attempted by immature students unless they are closely supervised. You might choose instead to turn these experiments into teacher demonstrations.

Unusual hazards are noted in the teaching notes and task cards where appropriate. But the curriculum cannot anticipate irresponsible behavior or negligence. It is ultimately the teacher's responsibility to see that common sense safety rules are followed at all times. Begin with these basic safety rules:

1. Eye Protection: Wear safety goggles when heating liquids or solids to high temperatures.
2. Poisons: Never taste anything unless told to do so.
3. Fire: Keep loose hair or clothing away from flames. Point test tubes which are heating away from your face and your neighbor's.
4. Glass Tubing: Don't force through stoppers. (The teacher should fit glass tubes to stoppers in advance, using a lubricant.)
5. Gas: Light the match first, before turning on the gas.

☐ Communicate your grading expectations.

Whatever your philosophy of grading, your students need to understand the standards you expect and how they will be assessed. Here is a grading scheme that counts individual effort, attitude and overall achievement. We think these 3 components deserve equal weight:

1. Pace (effort): Tally the number of check points you have initialed on the graph paper attached to each student's file folder or science notebook. Low ability students should be able to keep pace with gifted students, since write-ups are evaluated relative to individual performance standards. Students with absences or those who tend to work at a slow pace may (or may not) choose to overcome this disadvantage by assigning themselves more homework out of class.

2. Participation (attitude): This is a subjective grade assigned to reflect each student's attitude and class behavior. Active participators who work to capacity receive high marks. Inactive onlookers, who waste time in class and copy the results of others, receive low marks.

3. Exam (achievement): Task cards point toward generalizations that provide a base for hypothesizing and predicting. A final test over the entire module determines whether students understand relevant theory and can apply it in a predictive way.

Gathering Materials

Listed below is everything you'll need to teach this module. You already have many of these items. The rest are available from your supermarket, drugstore and hardware store. Laboratory supplies may be ordered through a science supply catalog.

Keep this classification key in mind as you review what's needed:

special in-a-box materials: Italic type suggests that these materials are unusual. Keep these specialty items in a separate box. After you finish teaching this module, label the box for storage and put it away, ready to use again the next time you teach this module.	general on-the-shelf materials: Normal type suggests that these materials are common. Keep these basics on shelves or in drawers that are readily accessible to your students. The next TOPS module you teach will likely utilize many of these same materials.
(substituted materials): Parentheses enclosing any item suggests a ready substitute. These alternatives may work just as well as the original, perhaps better. Don't be afraid to improvise, to make do with what you have.	*optional materials: An asterisk sets these items apart. They are nice to have, but you can easily live without them. They are probably not worth an extra trip to the store, unless you are gathering other materials as well.

Everything is listed in order of first use. Start gathering at the top of this list and work down. Ask students to bring recycled items from home. The teaching notes may occasionally suggest additional student activity under the heading "Extensions." Materials for these optional experiments are listed neither here nor in the teaching notes. Read the extension itself to find out what new materials, if any, are required.

Needed quantities depend on how many students you have, how you organize them into activity groups, and how you teach. Decide which of these 3 estimates best applies to you, then adjust quantities up or down as necessary:

$Q_1 / Q_2 / Q_3$

Single Student: Enough for 1 student to do all the experiments.
Individualized Approach: Enough for 30 students informally working in 10 lab groups, all self-paced.
Traditional Approach: Enough for 30 students, organized into 10 lab groups, all doing the same lesson.

KEY:	*special in-a-box materials* general on-the-shelf materials
	(substituted materials) *optional materials

1/10/10 rolls masking tape	1/10/10 eyedroppers	5/50/50 straws
1/10/10 scissors	1/10/20 lab thermometers	3/30/30 straight pins
1/2/2 *sheets pressed cardboard*	1/4/10 lids for baby food (small) jars	1/2/5 paper punches
4/25/25 *milk cartons — quart or half gallon sizes*	1/1/1 roll aluminum foil	1/4/10 margarine tubs (bowls)
	2/20/20 cups standard ice cubes	1/4/10 empty cans
2/2/2 *sets milk carton objects — see notes 2*	1/1/1 box table salt	1/4/10 magnifying glass
	1/1/1 roll paper towels	1/1/1 roll toilet tissue
1/5/10 *magnets	1/10/10 candles (may use Bunsen burners or alcohol lamps in some activities)	2/20/20 quart jars with lids
1/10/10 plastic produce bags		1/1/1 roll string
1/1/1 pkg popcorn, unpopped		1/1/1 *hot plate or radiator
1/1/2 dictionaries	1/10/10 pkgs matches	1/5/10 plastic syringes — 3 cc capacity, no needles
3/18/30 baby food jars (small jars)	1/10/10 wooden clothespins	
1/1/1 source hot/cold water	.4/4/4 *cups snow or ice shavings — see notes 7	1/10/10 *plastic 2 liter bottles with lids — see notes 16*
1/1/1 cider jug or equivalent		
1/1/1 bottle food coloring	.2/2/2 cups wax shavings — see notes 8	1/5/10 *twist ties (thin wire)*
1/1/1 wall clock with second hand (wristwatch or stopwatch)	1/4/10 *index cards — 4 x 6 inches	1/4/10 hand calculators
	2/10/20 styrofoam cups	1/2/5 rolls cellophane tape
.2/1/2 *meters narrow tubing, 1/8 inch diameter or less — see notes 5*	2/20/20 plastic sandwich bags	3/30/30 birthday candles
	1/1/1 box paper clips	2/20/20 size-D dry cells
2/20/20 test tubes - at least half should be large capacity	1/1/1 bottle rubbing alcohol	1/4/10 graduated cylinders — 100 ml capacity
	3/30/30 rubber bands	
.1/1/1 cup modeling clay (1-hole rubber stoppers)	1/1/1 roll plastic wrap	1/4/10 graduated cylinders — 10 ml capacity

D

Sequencing Task Cards

     This logic tree shows how all the task cards in this module tie together. In general, students begin at the trunk of the tree and work up through the related branches. As the diagram suggests, the way to upper level activities leads up from lower level activities.

     At the teacher's discretion, certain activities can be omitted or sequences changed to meet specific class needs. The only activities that must be completed in sequence are indicated by leaves that open *vertically* into the ones above them. In these cases the lower activity is a prerequisite to the upper.

     When possible, students should complete the task cards in the same sequence as numbered. If time is short, however, or certain students need to catch up, you can use the logic tree to identify concept-related *horizontal* activities. Some of these might be omitted since they serve only to reinforce learned concepts rather than introduce new ones.

     On the other hand, if students complete all the activities at a certain horizontal concept level, then experience difficulty at the next higher level, you might go back down the logic tree to have students repeat specific key activities for greater reinforcement.

     For whatever reason, when you wish to make sequence changes, you'll find this logic tree a valuable reference. Parentheses in the upper right corner of each task card allow you total flexibility. They are left blank so you can pencil in sequence numbers of your own choosing.

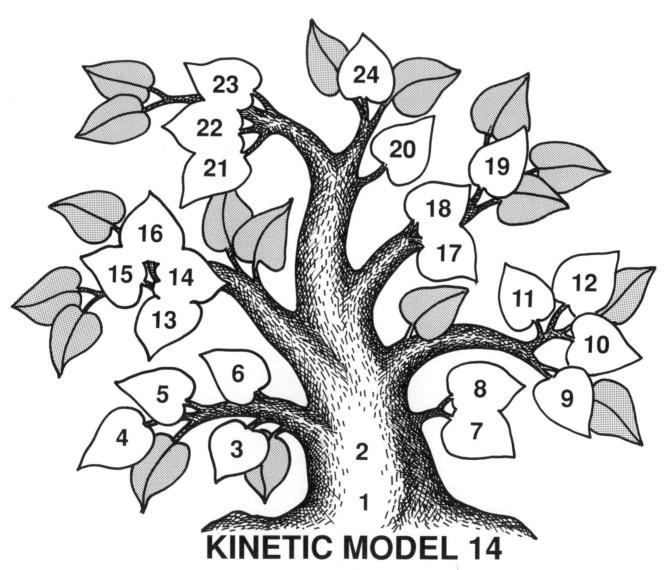

KINETIC MODEL 14

E

Long-Range
Objectives

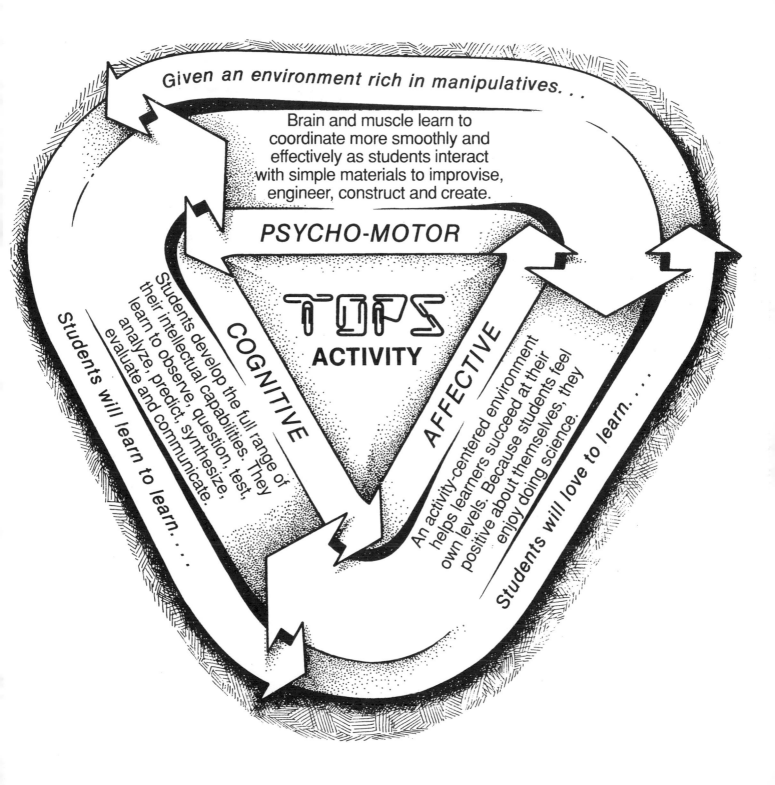

Given an environment rich in manipulatives. . .

Brain and muscle learn to coordinate more smoothly and effectively as students interact with simple materials to improvise, engineer, construct and create.

PSYCHO-MOTOR

TOPS ACTIVITY

COGNITIVE

Students develop the full range of their intellectual capabilities. They learn to observe, question, test, analyze, predict, synthesize, evaluate and communicate.

AFFECTIVE

An activity-centered environment helps learners succeed at their own levels. Because students feel positive about themselves, they enjoy doing science.

Students will learn to learn. . . .

Students will love to learn. . . .

Review / Test Questions

Photocopy the questions below. On a separate sheet of blank paper, cut and paste those boxes you want to use as test questions. Include questions of your own design, as well. Crowd all these questions onto a single page for students to answer on another paper, or leave space for student responses after each question, as you wish. Duplicate a class set and your custom-made test is ready to use. Use leftover questions as a review in preparation for the final exam.

task 1-2
A small child shakes the contents of a gift-wrapped box to try to guess what is inside. Is this child practicing science? Explain.

task 1-3
Devise a demonstration to show that air is a real substance even though it is invisible.

task 3-4
Devise an experiment using a bottle of perfume to show that perfectly still air is composed of molecules in constant motion.

task 3-5
Wet the rim of an empty glass beverage bottle, then lay a coin on top to seal it. Squeeze it tightly in your hands to make the coin jump. What is happening?

task 6-8 A
Identify 4 common phase changes in water. For each change, indicate if heat is taken in or given out.

task 6-8 B
Sulfur has a melting point of 113° C and a boiling point of 357° C.
a. Suppose you heat it to 200° C. Is it solid, liquid or gas at this temperature?
b. On the graph below, sketch how this compound cools over time if you remove it from the heat and allow it to cool to room temperature.

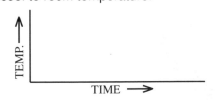

task 6-8 C
Water in a tea kettle boils at 100° C. Would touching the liquid burn you as much as touching the vapor? Explain.

task 7-8
Can you heat water without raising its temperature? Explain.

task 9
Never boil water in a sealed jar. Why is this dangerous?

task 10-13 A
Why does evaporation cool a liquid?

task 10-13 B
An empty glass takes on the same temperature as its surroundings.
a. Why is this not generally true if the glass is filled with water?
b. Under what conditions would this be true for a glass full of water? Explain.

task 11
You have a balloon and a bottle of perfume. Design an experiment to determine if the skin of your balloon is permeable to perfume molecules.

tasks 13, 14
A hygrometer hangs outside your house on a foggy day. The dry thermometer reads 10.0° C. What does the wet thermometer read? Why?

tasks 10, 14
Explain how a hygrometer measures moisture content in the air.

task 15-16 A
After hours of high speed driving across a hot desert road, you decide to pull off and check your tires. They look a little over-inflated, so you let out some air by pushing on each valve with your fingernail. Does the air feel warm? Explain.

task 15-16 B
Why does warm, moist air form clouds when it rises?

task 17-18
A wire is attached to a meter stick at the 1 cm mark and wrapped around the back of a chair to suspend the meter stick as shown.

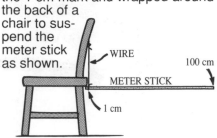

a. What happens to the meter stick when the wire is heated with a candle flame, then allowed to cool?
b. If the length of the wire changes 1 mm, how far does the end of the meter stick move? Use similar triangles to illustrate your answer.

task 18-19 A
Strips of brass and iron are riveted together, then heated over a flame. If brass expands and contracts faster than iron, what can you expect to see?

task 18-19 B
To remove a tight metal twist-off lid from a glass jar, it is sometimes helpful to place the lid under hot water. Why is this effective?

task 20
Ten grams of ice at 0° C is placed in a jar. If the heat of fusion for water is 80 cal/gram, how many calories of heat are required to heat the water to 20° C? Show your work.

task 21-23 A
A balloon that is inflated to a volume of 600 ml at 20° C shrinks or expands 2 ml for each 1° C change in temperature. What is absolute zero according to this data?

task 21-23 B
A Pyrex flask with a capacity of exactly 373 ml is heated to 100° C in an oven, then quickly inverted into an ice bath at 0° C. Exactly 100 ml of water is drawn into the flask.
a. What volume does the air in the flask occupy at 0° C?
b. Use this data to find the value of absolute zero.

task 24
Cooking recipes sometimes include special directions for high altitude locations. Why is this necessary?

Answers

task 1-2
Yes. The child is experimenting, attempting to match shapes that he or she is familiar with (or would like to receive) with the sounds and motions of unknown objects.

task 1-3
Many answers are possible: blow on a piece of paper or your finger; inflate a balloon; submerge an inverted bottle under water and observe that its insides remain dry.

task 3-4
Find a room where the air is perfectly calm. Open the bottle of perfume and remain completely still. Soon you will be able to smell its scent as the perfume molecules evaporate, mix with randomly moving molecules in the calm air, then get bumped and jostled into the air you inhale.

task 3-5
Air is not being squeezed out of the rigid glass bottle. Rather, the warmth of your hands is increasing the kinetic energy of the air molecules inside, causing them to move faster and thus occupy more volume. The coin jumps as some of this expanding air pushes out of the bottle.

task 6-8 A
Ice melts to water (heat in).
Water evaporates to vapor (heat in).
Vapor condenses to water (heat out).
Water freezes to ice (heat out).

task 6-8 B
a. At 200° C sulfur is a liquid. It has melted but not yet boiled away.

b.

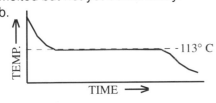

task 6-8 C
Touching the vapor would inflict a more serious burn, even though both phases have the same 100° C temperature. This is because additional heat energy would be released into your skin as the vapor condensed to water.

task 7-8
Yes. You can heat ice at 0° C to water at 0° C, or water at 100° C to vapor at 100° C.

task 9
Boiling water causes its liquid molecules to separate and occupy much more space as a gas. If the jar is sealed, pressure will build rapidly until the jar explodes.

task 10-13 A
Faster moving molecules in a liquid are more likely to evaporate than slower ones. The average kinetic energy of all the molecules left behind is lower. So the liquid is cooled.

task 10-13 B
a. A glass of water is generally cooler than its surroundings because it is cooled by evaporation.

b. If the relative humidity of the air above the glass of water was 100%, water molecules would condense into it (a warming effect) at the same rate they evaporated out of it (a cooling effect). The water would then assume the temperature of its surroundings. This would happen on a foggy day, or if the water was sealed in its container.

task 11
Add a drop or two of perfume to the inside of the balloon, being careful not to spill any on the outside. Then inflate the balloon and tie it off. If you can smell the perfume inside, then its molecules must be passing through the balloon skin. If there is no odor, try covering it in a bell jar or other container to test if a smell will accumulate over time. If not, the balloon skin is likely impermeable.

task 13-14
The wet thermometer reads 10.0° C also. It is not cooled at all by evaporation because foggy air is 100% saturated with water vapor.

task 10, 14
A hygrometer has two thermometers, one with a dry bulb and one that is wet. The dry thermometer measures the temperature of the air, while the wet one, cooled by evaporation, records a somewhat lower temperature. If this difference is large, evaporation is rapid because the air is dry. If it is small, evaporation is slow because the air is humid.

task 15-16 A
No. The air feels cool. Even though the high-pressure air inside the tires is warm, when released through the valve it expands to a much lower pressure outside the tire, losing its kinetic energy.

task 15-16 B
As warm air rises, it expands into lower pressures at higher altitudes, and therefore cools. As it cools, its relative humidity increases. Colliding water vapor molecules begin sticking together rather than bouncing off each other. Condensation droplets form to create clouds.

task 17-18
a. As the wire heats, the end of the meter stick dips down. As the wire cools, it rises back up.

b. Similar triangle ABC and AXY have proportional sides. If the wire changes 1 mm, changing XY by that amount, the end of the meter stick at BC moves 100 times more, or 10 cm.

task 18-19 A
On heating, the more rapidly expanding brass bends the strip inward around the iron. On cooling, the two metals resume their original lengths and the strip straightens.

task 18-19 B
Heat expands the metal lid more rapidly than the mouth of the glass jar. This increases the space between the two materials and loosens the lid.

task 20
Heat to melt ice at 0° C:

$$10 \text{ grams} \times \frac{80 \text{ cal}}{1 \text{ gram}} = 800 \text{ cal}$$

Heat to warm the water to 20° C:

$$\frac{20 \text{ cal}}{1 \text{ gram}} \times 10 \text{ grams} = 200 \text{ cal}$$

$$\text{total heat} = 1,000 \text{ cal}$$

task 21-23 A
If the balloon shrinks 2 ml for each drop in temperature of 1° C, then a loss of 600 ml (shrinking it to zero volume) requires a drop in temperature of 300° C. Thus,

$$\text{absolute zero} = 20° C - 300° C$$
$$= -280° C$$

task 21-23 B
a. volume of air at 100° C = 373 ml
– volume of water at 0° C = 100 ml
volume of air at 0° C = 273 ml

b. A volume reduction of 100 ml over a temperature drop of 100° C means that the air inside the flask loses 1 ml per 1° C. Its volume of 273 ml at 0° C, can cool another 273° C before collapsing to zero volume. Hence absolute zero must be -273° C.

task 24
Food must be cooked over longer periods of time at higher altitudes to compensate for the lower boiling temperature under lower atmospheric pressure.

TEACHING NOTES
For Activities 1-24

Task Objective (TO) devise ways to indirectly determine the shapes of cardboard patterns inside a milk carton.

INDIRECT EVIDENCE   O   Kinetic Model (   )

1. Tape the square, circle and triangle patterns to cardboard. Carefully cut out each shape.

MAKE THE CARDBOARD EDGES SMOOTH.

2. Put one shape into a clean, dry milk carton. Invent ways to identify it *without* touching it or looking at it directly.

☐ *or*
△ *or* ○

3. Refine your methods until you can correctly identify each shape that a friend hides inside. Explain your technique.

THE △ ALWAYS...

4. Scientists tell us that air is made from tiny *molecules* that are too small to see. How can they know about things they haven't seen?

© 1992 by TOPS Learning Systems   1

Answers / Notes

2. *This step does not involve guessing. Students already know what shape they put inside. The task here is to invent ways to turn and tilt the carton that will consistently discriminate among the possibilities.*

3. Turn the carton sideways and on edge so the figure slides into a "trough." Tilt this trough so the figure shifts to the carton's bottom and listen carefully. If it rolls it must be the circle. If it slides it could be the square or triangle. After it stops, slowly tip the carton into a vertical position and listen carefully. If it remains stationary it is the square. If it shifts to a new position it is the triangle.

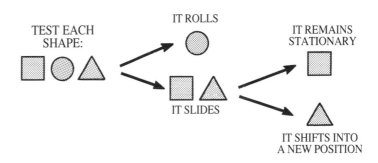

TEST EACH SHAPE:

IT ROLLS

IT SLIDES

IT REMAINS STATIONARY

IT SHIFTS INTO A NEW POSITION

4. As in this experiment, the nature of things unseen can be inferred indirectly by observation or experiment. The presence of unseen air molecules, for example, is easily detected as a cool breeze against the face or a breath of fresh air.

Materials

☐ Masking tape.
☐ Scissors heavy enough to cut cardboard.
☐ A square, circle and triangle pattern. Photocopy these from the supplementary page at the back of this book.
☐ Heavy pressed cardboard. Recycle a Grape-Nuts cereal box, or use the back of a writing tablet. Index cards and corrugated cardboard don't work as well.
☐ A waxed or plasticized paper milk carton. Quart or half gallon sizes are both suitable.

(TO) make inferences about the contents of a milk carton using indirect experimental evidence.

MODELING UNKNOWNS   Kinetic Model ( )

1. Get a sealed "mystery box." Experiment to determine the properties of the object inside, without unsealing your box.

a. Sketch the size and shape of each object inside.

b. Cite experimental evidence to support your drawing.

c. Compare the properties of each unknown object to known objects that you put inside a second milk carton.

2. Name a physical property of your object that you can't be sure about.

3. Repeat these steps with other mystery boxes coded with other letters.

4. To correctly model real-world science, is it fair to remove the tape from any mystery box and peek inside? Defend your answer.

© 1992 by TOPS Learning Systems   2

Answers / Notes

1. *Objects are coded by shared characteristics: (a) Disks that roll or slide. (b) Objects that pivot about a point. (c) Spheres that roll in all directions. (d) Long cylinders that roll and slide. (e) Other shapes.*

SAMPLE ANSWER: The object seems fairly heavy. It rolls easily from one side of the carton to the other. It slides a short distance from top to bottom. It is non-magnetic.

A candle, a piece of chalk and a test tube were placed one at a time in another milk carton and moved about. Only the test tube produced a similar glass-hitting-cardboard sound.

2. A variety of properties are difficult or impossible to discern through box manipulation alone: color (Is the marble blue?); internal structure (Does the washer have a hole?); material composition (Is the nail galvanized?).

3. *This activity may require a number of class periods, depending on the number of boxes students wish to test.*

4. No. Looking inside the box is like opening nature and understanding all there is to know. If this were possible, there would be no need to practice science.

Materials

☐ Sealed quart or half gallon milk cartons, labeled by code, that contain the following objects:

a1. washer	b1. tack	c1. marble	d1. half straw	e1. clothespin
a2. penny	b2. nail	c2. BB	d2. AA battery	e2. a 5-link paper clip chain
a3. bottle cap	b3. rubber stopper	c3. clay ball	d3. test tube	e3. microscope slide

Make up at least one mystery box per student, duplicating or omitting boxes according to class size. Save boxes with objects and codes intact to use again the next time you teach this module.

☐ Clean, dry, milk cartons, remaining open and empty, 1 per student.

☐ Additional objects for comparison testing. Use items in lists a-e above, plus miscellaneous chalk, buttons, candles, rubber bands, etc.

☐ Magnets (optional).

(TO) develop a kinetic model for air trapped in a plastic bag.

A KINETIC MODEL   O   Kinetic Model (   )

1. Fully open a plastic bag, then twist its mouth closed to form a balloon. What does this "mystery bag" contain? Describe its contents as fully as possible.

2. Set a jar of water next to a jar of air. If you magnified the molecules in each about twenty million times, they would have these relative sizes:

a. Identify the main elements in water and air molecules.

b. These molecules are nearly the same size, yet water is plainly visible, and air is not. Propose an explanation.

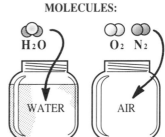

MOLECULES:
H_2O O_2 N_2
WATER   AIR

3. Let kernels of popcorn model air molecules. Vigorously shake a handful of these "molecules" inside the plastic balloon from step 1.

a. What properties of air does this model help to explain?

b. Why call this popcorn representation a *kinetic* model?

c. Does this kinetic model suggest that all air molecules move at the same speed? Observe closely.

d. Does this model fail to act like an actual bag of air in any respect?

3

Answers / Notes

1. This bag is filled with air — an invisible, lightweight, compressible substance that occupies space.

2a. Water molecules are made from hydrogen *and* oxygen atoms. Air molecules are made from oxygen *or* nitrogen atoms. *(Air also contains argon, carbon dioxide and trace amounts of many other gases.)*

2b. Water is visible because its molecules are packed close enough together to be seen in mass. Air molecules, by contrast, are widely separated from each other. They are far too small to be seen individually.

3a. This model helps to explain how air molecules maintain their separation one from another, thus remaining invisible, lightweight and compressible. It further suggests why the sides of the bag remain pushed out. Air molecules beat against the plastic and collide with each other, maintaining a steady outward pressure.

3b. The word "kinetic" refers to movement. The model suggests moving molecules.

3c. The "molecules" move at many different speeds depending on how they collide with the container and each other. Close observations reveals that some even stop, momentarily frozen in space. *(At room temperature an air molecules might be moving from near 0 mph to speeds in excess of 2000 mph. Taken as a whole, the average molecular speed is about 1000 mph.)*

3d. Unlike this crude kinetic model, a real bagful of air maintains its shape without being continuously shaken. *(This suggests that air molecule collisions are energy conserving, or perfectly elastic.)*

Materials

☐ A plastic produce bag.
☐ Popcorn, unpopped.
☐ A dictionary.
☐ Two baby food jars.
☐ Water.

(TO) show by indirect experimental evidence that water molecules are in continuous motion.

KINETIC MIX? ◯ Kinetic Model ( )

1. Fill 3 small jars with water that has come to room temperature. Set them on a solid surface, free of vibrations, so the water becomes very calm.

NOW: IN 2 MINUTES: IN ANOTHER 2 MINUTES:

FIRST JAR SECOND JAR THIRD JAR

2. Add 1 drop of food coloring to the first jar. At 2 minute intervals, repeat for the second and third jars. Each drop should cleanly separate from the dropper *before* it touches the water.

    a. Describe the color patterns that form in the water. How do they change over time?
    b. Do you think the kinetic model for air applies to water? Why?
    c. How might these jars look after 24 hours? Make a prediction.

3. Test your prediction with a fresh jar of calm water resting in a safe, out-of-the-way place.

4

Answers / Notes

2a. The drop of food coloring leaves a complex, looping trail as it falls through the water. Sometimes it forms a ring. Initially these patterns have sharply defined edges, but over time these edges begin to blur.

2b. The kinetic model for air seems to apply to liquid water as well. The progressively blurring edges of the color patterns suggest that moving dye molecules are gradually colliding with and mixing into moving water molecules.

2c. If this mixing process continues, the food coloring could well mix throughout the entire jar of water, creating a homogeneous color.

3. After just 6 hours the food coloring is uniformly distributed throughout the water.

Materials

☐ Three baby food jars.
☐ A closed jug of water labeled "room-temperature water". Tap water is also suitable, as long as temperature variations from one jar to the next are not extreme. A special closed water jug should be prepared now, however, since it will be required throughout this module.
☐ Food coloring. Blue is a dark, intense color that is easy to observe. If it doesn't come suitably packaged, dispense this in a labeled dropper bottle.
☐ A wall clock.

(TO) build an air thermometer that is sensitive to small temperature changes. To explain how it functions in terms of the kinetic model.

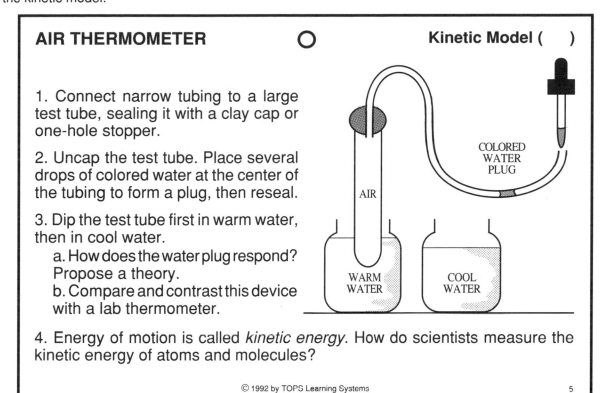

AIR THERMOMETER ◯ **Kinetic Model ( )**

1. Connect narrow tubing to a large test tube, sealing it with a clay cap or one-hole stopper.

2. Uncap the test tube. Place several drops of colored water at the center of the tubing to form a plug, then reseal.

3. Dip the test tube first in warm water, then in cool water.
   a. How does the water plug respond? Propose a theory.
   b. Compare and contrast this device with a lab thermometer.

4. Energy of motion is called *kinetic energy*. How do scientists measure the kinetic energy of atoms and molecules?

COLORED WATER PLUG

AIR

WARM WATER

COOL WATER

5

Answers / Notes

1. *One way to fashion a clay cap is to first roll it into a ball. Place it on the mouth of the test tube, then punch a pencil point through its center.*

2. *The illustrated eyedropper suggests only one of several ways to place colored water inside the tube. The end of the tube can also be dipped into the water, then sealed at the other end with the thumb and withdrawn. Capillary action also draws water into very fine diameter tubing.*

3a. The water plug is extremely sensitive to small temperature changes inside the test tube. In warm water it is pushed outward, suggesting that the air inside the tube expands as it warms. In cool water the plug is drawn inward, suggesting that the air contracts as it cools.

3b. A lab thermometer containing mercury or red alcohol functions in a similar manner. As its tiny bulb warms up or cools down, the liquid inside moves up and down its glass column to show a change in temperature. Air in the test tube expands and contracts to a much greater degree, making it a more sensitive, though less practical, instrument.

4. Scientists measure the kinetic energy of atoms and molecules by taking their temperature. The hotter (or cooler) something is, the faster (or slower) its atoms and molecules move.

Materials

☐ About 20 cm of narrow tubing with an inside diameter of 1/8 inch (3 mm) or less. Larger diameters will not work. Use clear plastic tubing if available. Glass or natural rubber tubing (not black) will also serve. Don't overlook recycled tubing from empty pump bottles or used ballpoint pen fillers.
☐ A test tube. Large capacity tubes give the most dramatic results.

☐ Oil-based modeling clay, or a 1-hole rubber stopper sized to fit the test tube. If you elect ot use stoppers *and* narrow glass tubing, fit them together in advance to avoid accidents.
☐ A jar of water tinted by a drop of food coloring. If your tubing is clear, tinting is unnecessary.
☐ An eyedropper (optional).
☐ Two baby food jars or equivalent.
☐ Hot and cold tap water.
☐ A lab thermometer.

(TO) observe and describe four phase changes in water. To interpret these changes on a molecular level.

PHASE CHANGES   ◯   Kinetic Model ( )

1. Water exists in 3 *phases* — solid (ice), liquid (water) and gas (vapor). Fully describe each *phase change* below using the capitalized vocabulary.

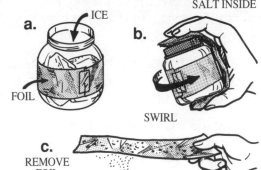

a. CONDENSE: Tape a strip of foil around a small jar, then drop 2 ice cubes inside. Observe the foil.

b. FREEZE: Add just enough salt to bury the ice, then seal with a lid. Swirl this around the wall of the jar for several minutes, holding it top and bottom with a paper towel.

c. MELT: Remove the frosted foil from the jar, while keeping it under close observation.

d. EVAPORATE: Lightly warm the water droplets on the foil over a flame.

2. During which phase changes do water molecules move faster? Slower?

3. During which phase changes, therefore, must water molecules gain heat? Release heat?

6

Answers / Notes

1a. Water vapor in the air *condenses* on the cold foil into tiny liquid water drops.

1b. The liquid water drops *freeze* on the cold foil into tiny lumps of ice. Additional water vapor in the air freezes on contact with the foil, forming a white layer of frost. *(This frost buildup can be dramatic under humid conditions.)*

1c. The white frost and ice quickly melt to transparent water drops as the foil is separated from its cold jar.

1d. The water drops quickly *evaporate* into vapor, disappearing from the foil entirely, leaving behind tiny deposits of dissolved minerals. *(Insist that your students describe gaseous water as vapor, not steam. Steam, like clouds or fog, consists of tiny drops of liquid water that have already condensed. Water molecules in gas form are individuated, and hence, totally invisible.)*

2. Water molecules move faster when they…
     …melt from ice into water.
     …evaporate from water into vapor.
    Water molecules move slower when they…
     …condense from vapor into water.
     …freeze from water into ice.

3. Water molecules absorb heat as they melt and as they evaporate.
    Water molecules lose heat as they condense and as they freeze.

Materials

☐ A baby food jar and lid.
☐ Aluminum foil.
☐ Scissors.
☐ Masking tape.
☐ Two ice cubes that are small enough to fit inside the jar. Cubes from a standard refrigerator ice tray are appropriate.

☐ Table salt.
☐ A paper towel.
☐ Any heat source plus matches. A candle, alcohol lamp, Bunsen burner, hot plate or heat radiator all serve.

(TO) plot a heating curve for water. To recognize that temperature remains constant as water changes phase.

HEATING CURVE ○ Kinetic Model ( )

1. Clamp a clothespin handle near the mouth of a large test tube. Pry it wider, as necessary, to fit around the tube.

2. Fill the tube with snow or ice shavings. Insert a thermometer to the bottom to record the lowest temperature. Call this time zero.

3. Immediately heat the ice over a candle flame while stirring with a thermometer. Ask a friend to record the temperature every 30 seconds in a data table. Continue until the water has boiled for 2 minutes.

4. Plot your data on a graph. Circle each point, then connect them with a smooth, curving line.

5. Does the temperature of water always rise as water absorbs heat? Explain.

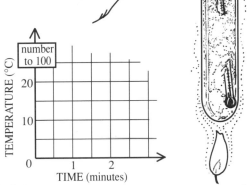

TIME (min)	TEMP. (°C)
0	
0.5	
1.0	
1.5	
.	
.	
.	

© 1992 by TOPS Learning Systems 7

Answers / Notes

3.

TIME (min)	TEMP. (°C)
0	1
0.5	1
1.0	2
1.5	11
2.0	23
2.5	36
3.0	46
3.5	53
4.0	60
4.5	72
5.0	78
5.5	82
6.0	88
6.5	96
7.0	100
7.5	100
8.0	100

4.
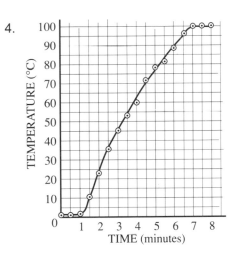

5. No. The heating curve is flat at the bottom (where ice melts to water) and at the top (where water evaporates to steam). This shows that heat from the candle flame is used to drive each phase change *without* a corresponding rise in temperature. Only through the ascending central portion of the curve is most of the candle heat used to raise the water's temperature.

Materials

☐ A jar of snow. If the season or your geographic location is wrong, you can make a suitable ice slush with ice cubes and a food blender. Or you can manufacture snow using a cheese grater. Wrap the end of each ice cube in a thick rubber band to provide a grip. You (or your students) will only be able to shave about half the ice cube before it gets too small to handle, but this should provide enough ice shavings to fill at least one large test tube. Do not substitute chipped ice unless the pieces are small enough to fit into your test tube beside the thermometer.

☐ A clothespin.
☐ A large test tube.
☐ A thermometer.
☐ A candle and matches. A hotter alcohol lamp or a Bunsen burner will melt the ice too rapidly.
☐ A wristwatch, stopwatch, or wall clock with a second hand.
☐ Graph paper. Photocopy this from the supplementary page.

(TO) plot a cooling curve for candle wax. To recognize that heat is lost as wax undergoes a phase change from liquid to solid.

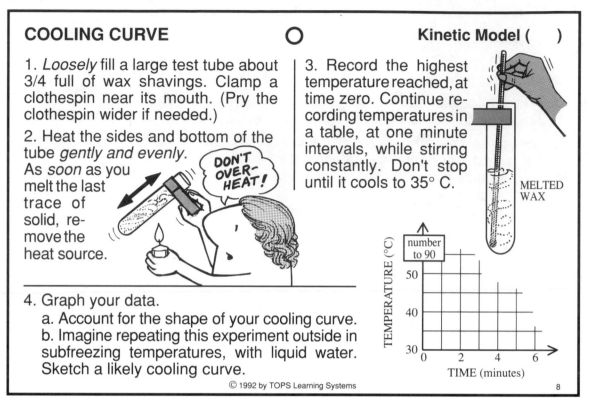

COOLING CURVE   ○   Kinetic Model (   )

1. *Loosely* fill a large test tube about 3/4 full of wax shavings. Clamp a clothespin near its mouth. (Pry the clothespin wider if needed.)

2. Heat the sides and bottom of the tube *gently and evenly*. As *soon* as you melt the last trace of solid, remove the heat source.

DON'T OVER-HEAT!

3. Record the highest temperature reached, at time zero. Continue recording temperatures in a table, at one minute intervals, while stirring constantly. Don't stop until it cools to 35° C.

MELTED WAX

4. Graph your data.
   a. Account for the shape of your cooling curve.
   b. Imagine repeating this experiment outside in subfreezing temperatures, with liquid water. Sketch a likely cooling curve.

© 1992 by TOPS Learning Systems   8

Answers / Notes

1-3. *A test tube 3/4 filled with wax shavings should yield about 1/4 tube of melted wax. At normal room temperature, this volume of wax will require perhaps 25 minutes to cool below 35° C. Supervise your students closely. Under NO circumstances should they continue heating the liquid paraffin after all the solid has melted: the liquid will heat far beyond 100°, and the resulting vapors could possibly ignite.*

3.

TIME (min)	TEMP. (°C)
0	79.0
1	70.0
2	63.0
3	57.5
4	54.0
5	53.0
6	53.0
7	52.7
8	52.1
9	52.0
10	51.8
11	51.2
12	50.8
13	50.3
14	49.9
15	49.4
16	48.4
17	47.0
18	44.5
19	41.6
20	39.0
21	37.0
22	35.8
23	34.3

4.

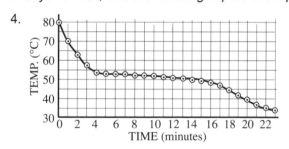

4a. The cooling curve begins as an ever-moderating downward slope: the liquid wax loses heat less rapidly as the temperature difference with its surroundings decreases. At about 53° C this curve flattens abruptly: though the temperature remains nearly constant, heat is still lost as wax changes phase from liquid to solid. At about 49° C the moderating downward slope resumes: the now solid wax is cooling toward room temperature.

Candle wax is a complex mixture of hydrocarbons with slightly different melting points. If it were a pure substance with just one melting point, the phase change portion of this graph would flatten completely.

4b.

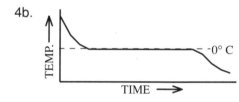

Materials

☐ Wax shavings. Scrape these with scissors from candles or canning paraffin sold in grocery stores. You may wish to prepare these ahead. In well-ventilated rooms, you may substitute moth balls.

☐ A large test tube and a clothespin.

☐ A candle or other heating source plus matches.

☐ A thermometer.

☐ A watch, clock or stopwatch with a second hand.

☐ Graph paper. Photocopy this from the supplementary page.

(TO) evaporate water in closed systems. To use the kinetic model to account for its dramatic increase in volume.

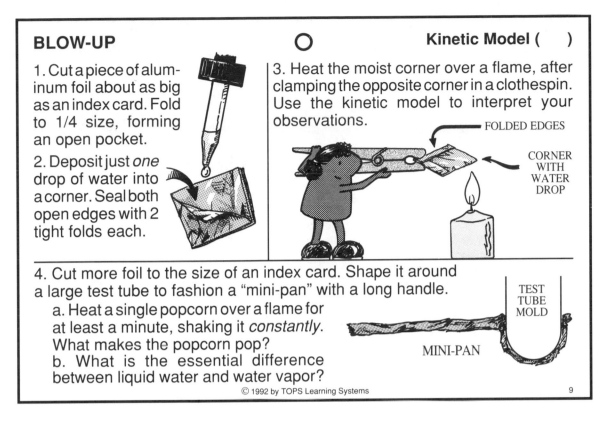

BLOW-UP   ○   **Kinetic Model (   )**

1. Cut a piece of aluminum foil about as big as an index card. Fold to 1/4 size, forming an open pocket.

2. Deposit just *one* drop of water into a corner. Seal both open edges with 2 tight folds each.

3. Heat the moist corner over a flame, after clamping the opposite corner in a clothespin. Use the kinetic model to interpret your observations.

FOLDED EDGES

CORNER WITH WATER DROP

4. Cut more foil to the size of an index card. Shape it around a large test tube to fashion a "mini-pan" with a long handle.

a. Heat a single popcorn over a flame for at least a minute, shaking it *constantly*. What makes the popcorn pop?

b. What is the essential difference between liquid water and water vapor?

TEST TUBE MOLD

MINI-PAN

9

Answers / Notes

3. The flat foil pocket puffs fully out. Heating the water drop causes its molecules to move faster and faster until they evaporate in great numbers into the gas phase. These vapor molecules separate widely from each other to occupy a much larger volume. They beat against the sides of the foil pocket and force it to bulge out.

4a. Evaporating water makes popcorn pop. Water molecules inside the kernel move more energetically as they are heated, evaporating quickly from liquid to gas. This expands the corn seed, literally blowing it open.

4b. Water molecules in the gas phase move faster and take up more space. They bounce about vigorously and independently. They don't cling together as in the liquid phase.

Materials

☐ Aluminum foil.
☐ Scissors.
☐ A 4 x 6 index card (optional).
☐ An eyedropper.
☐ Water.
☐ A candle, alcohol lamp or Bunsen burner.
☐ Matches.
☐ A clothespin.
☐ A large test tube.
☐ Popcorn.

(TO) understand why liquids cool as they evaporate.

COOL IT! ○ Kinetic Model (   )

CLOSED BAG: OPEN BAG:

ROOM TEMPERATURE WATER

1. Line 2 cups with plastic sandwich bags, then fill each half full with room temperature water.

2. Twist and paper-clip one bag closed; leave the other open. Over a long period of time, would you expect to observe changes in either bag? Explain.

3. Push the nose of a thermometer deep into the side of each bag, taking the temperature of its water *through* the plastic. Record your results to the nearest 0.1° C.

4. As water evaporates, which molecules are more likely to move into the gas phase — slow moving molecules or fast moving ones? Why?

5. Reconcile your answer in step 4 with your experimental result in step 3.

10

Answers / Notes

2. Over time, the open bag of water will dry up, its water evaporating out of the bag and mixing with the air as vapor. Water inside the closed bag, by contrast, should remain trapped inside.

3. Depending on your local humidity, students should report temperatures with a difference of perhaps one or two degrees. The closed bag should always be warmer than the open bag.

4. The faster moving water molecules are more likely to evaporate. They possess sufficient kinetic energy to break free from their neighbors at the surface of the liquid and escape as a gas.

5. Evaporation is a cooling process: faster moving molecules escape into the gas phase, leaving less energetic, slower moving molecules behind. These slower molecules have a slightly lower average kinetic energy, and therefore a slightly cooler temperature. Since evaporation occurs mainly in the open bag, its water is comparatively cooler than water in the closed bag.

Some molecules do evaporate in the closed bag as well, escaping as vapor into the closed air space above. There is little cooling, however, since vapor molecules in this closed atmosphere also warm the water by condensing back into the liquid phase. At equilibrium, cooling evaporation is exactly offset by warming condensation. There is no net temperature change. Your class will explore this idea in activity 13.

$$\text{WATER} + \text{HEAT} \underset{\text{condenses}}{\overset{\text{evaporates}}{\rightleftharpoons}} \text{VAPOR}$$

Materials

□ Two drinking cups. Paper cups or styrofoam coffee cups can both be used in this activity, but only styrofoam is suitable for activity 20.

□ Two plastic sandwich bags.

□ A *closed* jug of water labeled "room-temperature water." This must be prepared in advance to allow it enough time to reach ambient temperature.

□ A paper clip.

□ A thermometer.

(TO) observe and explain the cooling effects of alcohol. To trace the movement of alcohol molecules through a permeable membrane.

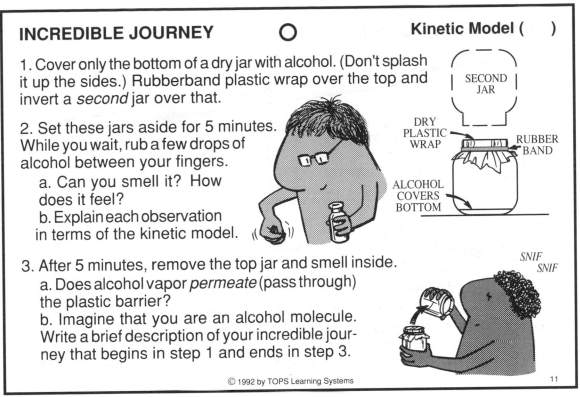

INCREDIBLE JOURNEY ○   **Kinetic Model ( )**

1. Cover only the bottom of a dry jar with alcohol. (Don't splash it up the sides.) Rubberband plastic wrap over the top and invert a *second* jar over that.

2. Set these jars aside for 5 minutes. While you wait, rub a few drops of alcohol between your fingers.
   a. Can you smell it? How does it feel?
   b. Explain each observation in terms of the kinetic model.

3. After 5 minutes, remove the top jar and smell inside.
   a. Does alcohol vapor *permeate* (pass through) the plastic barrier?
   b. Imagine that you are an alcohol molecule. Write a brief description of your incredible journey that begins in step 1 and ends in step 3.

SECOND JAR

DRY PLASTIC WRAP   RUBBER BAND

ALCOHOL COVERS BOTTOM

SNIF SNIF

© 1992 by TOPS Learning Systems   11

Answers / Notes

2a. The liquid has a strong characteristic odor. It makes the fingers feel remarkably cool.

2b. Alcohol molecules on your fingers move about continuously in the liquid phase. Those at the surface, moving fast enough and in the right direction, break free of their neighbors to evaporate as gas in sufficient numbers to detect by smell. The liquid molecules that remain on your fingers now feel cooler, because they lacked sufficient kinetic energy (a high enough temperature) to evaporate as rapidly as their faster moving neighbors.

3a. After 5 minutes, enough vapor permeates the plastic wrap to create a definite alcohol odor in the top jar.

3b. Hello. I am an alcohol molecule. You can call me Alco. I am presently flying solo just above a bookcase at the back of the room. Nitrogen and oxygen molecules occasionally whiz by. Sometimes we collide. Only rarely do I bump into other molecules like myself. Let me tell you of my incredible journey over the last 5 minutes.

   I was poured into a jar, along with billions upon billions of my kind. We jostled and tumbled about, shoulder to shoulder, until suddenly I broke loose and evaporated above the maddening crowd. I flew randomly about, bumping into other air and alcohol molecules that were as energetic as myself. I worked my way up, higher and higher, until I bumped into plastic wrap that covered the top of the jar. Somehow I managed to miss bouncing off the closely spaced plastic molecules that composed this barrier, zipping instead through its intermolecular spaces. After permeating this barrier, I continued up into the second jar with just a few friends, leaving many others that rebounded far below.

   I remained several more minutes in this top jar until something very strange happened. A giant nose inhaled me into this crazy olfactory receptor, then just as suddenly breathed me back out. I eventually worked my way over to the bookcase where I began my story. And now I'm wondering what will happen next!

Materials

☐ Two dry baby food jars.
☐ Rubbing alcohol. If you dispense in your own dropper bottles, be sure the labels read "Poison."
☐ A rubber band.

☐ Plastic wrap.
☐ Scissors.
☐ A wall clock or wristwatch.

(TO) compare rates of evaporation in water and rubbing alcohol on a balance beam. To account for observed differences in terms of intermolecular bond strengths.

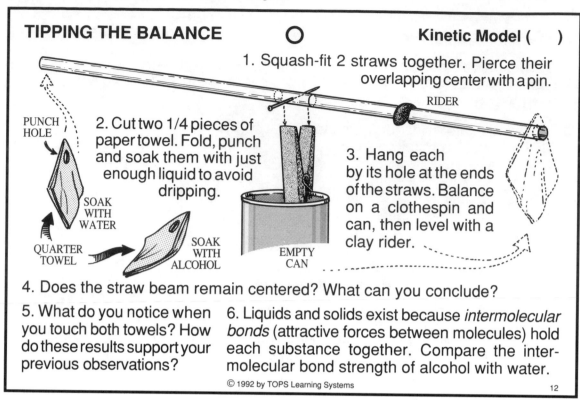

TIPPING THE BALANCE ○ **Kinetic Model ( )**

1. Squash-fit 2 straws together. Pierce their overlapping center with a pin.

RIDER

PUNCH HOLE

2. Cut two 1/4 pieces of paper towel. Fold, punch and soak them with just enough liquid to avoid dripping.

SOAK WITH WATER

QUARTER TOWEL

SOAK WITH ALCOHOL

EMPTY CAN

3. Hang each by its hole at the ends of the straws. Balance on a clothespin and can, then level with a clay rider.

4. Does the straw beam remain centered? What can you conclude?

5. What do you notice when you touch both towels? How do these results support your previous observations?

6. Liquids and solids exist because *intermolecular bonds* (attractive forces between molecules) hold each substance together. Compare the inter-molecular bond strength of alcohol with water.

© 1992 by TOPS Learning Systems

12

Answers / Notes

4. The straw beam tilts continuously up on the alcohol side, down on the water side. This happens fast enough to observe actual movement in the beam, demonstrating that alcohol loses weight by evaporation more rapidly than water.

You might point out to your more advanced students that relative losses by weight say nothing certain about the actual numbers of molecules that evaporate. Though alcohol molecules do evaporate in greater numbers, proof of this is left to the extension below.

5. The towel containing rubbing alcohol is much cooler than the towel containing pure water. This confirms that rubbing alcohol evaporates more rapidly, leaving behind liquid molecules with less kinetic energy.

6. The intermolecular bond strength of alcohol must be significantly weaker. Its bonds break more easily than bonds between water molecules, allowing alcohol molecules to evaporate more rapidly.

Extension

(1) Repeat this experiment with a gram balance. Compute evaporation losses for each liquid in grams/minute.
(2) A molecule of ethyl alcohol is about 2.6 times heavier than a water molecule; a molecule of isopropyl alcohol is about 3.3 times heavier. Molecule for molecule, does alcohol still evaporate faster than water? Show your work.

Materials

☐ Two straws.
☐ A straight pin.
☐ A paper towel.
☐ Scissors.
☐ A paper punch.
☐ Rubbing alcohol.

☐ Water.
☐ Two shallow bowls or small jars for soaking the pieces of paper towel.
☐ An empty can.
☐ A clothespin.
☐ Modeling clay.

(TO) condense water vapor against surfaces that are cooled by evaporation. To understand that evaporation and condensation form a dynamic equilibrium when humidity reaches 100%.

RELATIVE HUMIDITY　　　○　　　　Kinetic Model (　)

1. Add just enough water to a small jar so its bottom is fully covered. Cover tightly with a square of *dry* plastic wrap, then secure with a rubber band.

2. Deposit a drop of water on the left half of the plastic, then quickly put a drop of alcohol on the right.

3. Look at each drop with a magnifying lens. What begins to happen underneath? How fast?

4. Fan both drops about 200 times with a folded sheet of paper, then wipe them dry with a tissue.

　　a. Why are water vapor molecules more likely to condense under the drops than elsewhere on the plastic wrap? Why under one specific drop?

　　b. Does the area under each drop show *uniform* cooling? Why?

5. *Relative humidity* measures the capacity of air to hold water vapor. If the jar air has a relative humidity of 100% this means no additional water can evaporate into it without an equal amount also condensing back out.

　　a. Why do your droplets remain the same size?

　　b. How could you make them disappear?

© 1992 by TOPS Learning Systems　　　　　　　　13

Answers / Notes

3. Tiny drops of water condense under each drop, clouding the inside surface of the plastic wrap. This happens more rapidly under the drop of alcohol than under the drop of water.

4. *This step insures that each circle of condensation is surrounded with a ring of obviously larger droplets. Fanning may not be necessary for drops that have been left in place at least 5 minutes.*

4a. Molecules of water vapor condense as they cool and move more slowly. This happens on the coldest parts of the plastic, directly under each drop that is cooled by evaporation. More condensation accumulates under the alcohol drop because it evaporates most rapidly.

4b. No. The area under each drop appears cooler around its perimeter, as evidenced by its outside ring of larger condensation droplets. This is because evaporation happens only at the surface of the drop, not deep inside.

5a. The droplets remain unchanged over time because the jar has 100% humidity. That is, its inside atmosphere is fully saturated with water vapor. Molecules that evaporate out of these tiny drops into the saturated air around them displace equal numbers of vapor molecules that condense back into these same drops. This dynamic equilibrium creates no net change in drop size.

5b. To make the condensed droplets disappear, expose them to less saturated room air. Over time, more water will evaporate into this drier air than will condense back into the droplets. They will diminish in size and finally disappear.

Materials

☐ A baby food jar with water in the bottom.
☐ Plastic wrap.
☐ Scissors.
☐ A rubber band.

☐ A labeled dropper bottle of rubbing alcohol.
☐ A labeled dropper bottle of water.
☐ A magnifying lens.
☐ A piece of tissue.

(TO) construct a hygrometer. To measure the relative humidity of room air and compare it to other sources.

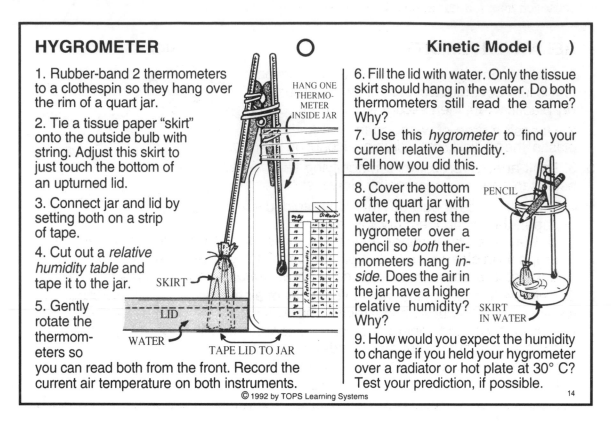

HYGROMETER ○   Kinetic Model ( )

1. Rubber-band 2 thermometers to a clothespin so they hang over the rim of a quart jar.

2. Tie a tissue paper "skirt" onto the outside bulb with string. Adjust this skirt to just touch the bottom of an upturned lid.

3. Connect jar and lid by setting both on a strip of tape.

4. Cut out a *relative humidity table* and tape it to the jar.

5. Gently rotate the thermometers so you can read both from the front. Record the current air temperature on both instruments.

HANG ONE THERMO-METER INSIDE JAR

SKIRT

LID
WATER
TAPE LID TO JAR

6. Fill the lid with water. Only the tissue skirt should hang in the water. Do both thermometers still read the same? Why?

7. Use this *hygrometer* to find your current relative humidity. Tell how you did this.

8. Cover the bottom of the quart jar with water, then rest the hygrometer over a pencil so *both* thermometers hang *inside*. Does the air in the jar have a higher relative humidity? Why?

9. How would you expect the humidity to change if you held your hygrometer over a radiator or hot plate at 30° C? Test your prediction, if possible.

PENCIL

SKIRT IN WATER

© 1992 by TOPS Learning Systems   14

Answers / Notes

1. *Wind the rubber band around the clothespin and thermometers, looping over a "jaw" of the clothespin to start, and around its "wing" to finish.*

5. *Students should give readings that agree to within 1° C of each other. (If they don't, decide which instrument gives the correct reading, then add or subtract a correction factor when reading the other.)*

6. The thermometers no longer read the same. The wet bulb now reads lower than the dry bulb because it is cooled by evaporation.

7. To report their current relative humidity, students should read their current room temperature on the dry bulb, then note how many degrees cooler the wet bulb is. These figures define a row and column that intersect on the relative humidity table. *(The wet bulb on the hygrometer may require several minutes to reach its final equilibrium temperature, both here and in step 8.)*

8. Yes. Evaporating water in the bottom of the jar keeps the inside air relatively moist. (If the air in your room is already close to 100% humidity, differences inside and outside the jar may not be significant.)

9. Heating the air has a drying effect. Its relative humidity should thus be significantly lower than room humidity. *(Students should hold the hygrometer just high enough above the heat source to read a dry temperature of 30° C. They must not touch the thermometer bulbs directly to the hot element. The mercury or red alcohol inside could easily expand beyond the capacity of the thermometer and damage the instrument.)*

Materials

☐ Two thermometers. Pair any that read excessively high, or low, for best possible agreement.
☐ A rubber band.
☐ A clothespin.
☐ A quart jar.
☐ Toilet tissue.
☐ String.
☐ Scissors.

☐ A jar lid. The lid from the quart jar is ideal.
☐ Masking tape.
☐ A table of relative humidity. Photocopy this from the supplementary sheet at the back of this module.
☐ Water at room temperature.
☐ A hot plate or radiator (optional). Never substitute an open flame.

(TO) investigate the relationship between the pressure on a gas and its temperature.

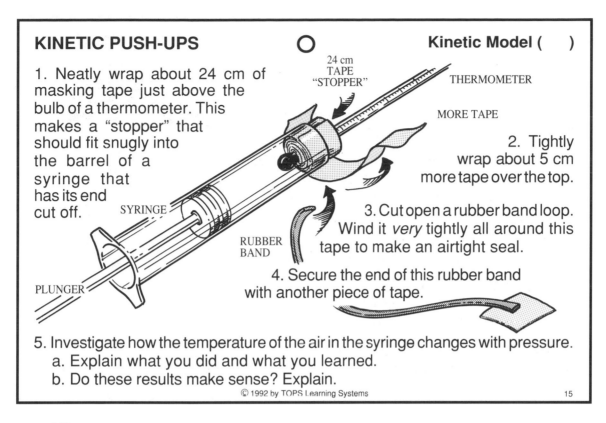

KINETIC PUSH-UPS ○ **Kinetic Model ( )**

1. Neatly wrap about 24 cm of masking tape just above the bulb of a thermometer. This makes a "stopper" that should fit snugly into the barrel of a syringe that has its end cut off.

24 cm TAPE "STOPPER"

THERMOMETER

MORE TAPE

2. Tightly wrap about 5 cm more tape over the top.

3. Cut open a rubber band loop. Wind it *very* tightly all around this tape to make an airtight seal.

SYRINGE

RUBBER BAND

PLUNGER

4. Secure the end of this rubber band with another piece of tape.

5. Investigate how the temperature of the air in the syringe changes with pressure.
   a. Explain what you did and what you learned.
   b. Do these results make sense? Explain.

© 1992 by TOPS Learning Systems   15

Answers / Notes

1. *In our experience, a standard lab thermometer requires about 24 cm of masking tape to enlarge it to the inside diameter of a 3 cc syringe. Substitute another length of masking tape in these directions if your equipment is different.*

2-3. *It is important to use enough tape to fully encircle the syringe and thermometer, yet not so much that it becomes inflexible. The rubber band, wrapped noose-like around the outside, will mold this tape around the end of the syringe, forming an airtight seal.*

5a. Pop out the plunger to fill the barrel with air. Then reinsert it and compress this air toward the thermometer bulb. As its pressure increases, the air heats perhaps 1° C. Relax the plunger to relieve this pressure, and the air temperature falls back about 1° C.

In summary, air temperature increases with increasing pressure and decreases with decreasing pressure.

5b. Yes. As the rubber plunger advances, compressing the air inside, molecules that strike it rebound with greater kinetic energy (a higher temperature) than they had before. Similarly, as the rubber plunger recedes, expanding the air inside, molecules that strike it rebound with less kinetic energy (a lower temperature) than they had before.

Materials

□ Masking tape and scissors.
□ A metric ruler. Photocopy this from the supplementary page.
□ A plastic 3 cc syringe that has its end cleanly removed. Cut this off in advance with good scissors, preserving the full length of the barrel.

□ A thermometer. Its bulb should fit inside the barrel of the syringe with a small clearance around its circumference.
□ A thin rubber band.

(TO) make a cloud in a bottle. To understand how pressure, temperature, and relative humidity affect cloud formation.

CLOUDS ◯ Kinetic Model (   )

LID

WATER

MATCH

SQUEEZE AND RELEASE

1. Wet the inside of a 2 liter plastic bottle with water, then drain the excess.

2. Light a match and drop it inside to create a little smoke. Seal airtight with a lid.

3. Make a cloud: alternately squeeze the bottle very tightly, then relax your grip. How can you make a cloud disappear, then reappear?

4. When water vapor condenses on tiny particles in smoke, it forms a cloud. How do the following variables work together to make this happen?

PRESSURE
TEMPERATURE
RELATIVE HUMIDITY

5. A steady breeze blows off the sea and over a range of mountains. Clouds gather at higher elevations but not below. What is happening?

16

Introduction

We live in an "ocean" of air in which pressure changes with depth. What happens as we move upward through this ocean in an elevator, car, or airplane? *(The pressure decreases. Higher pressure air builds up inside our ears. It eventually flows out as our ears "pop" to equalize the pressure).* What happens as we descend? *(The pressure increases. We can feel air outside our ears pressing against our eardrums until it enters with a popping sound.)*

Answers / Notes

3. The cloud vanishes each time you squeeze the bottle. It recondenses each time you relax your grip. *(The visibility of this cloud dramatically increases when spotlighted in a darkened room.)*

4. Squeezing the bottle increases the pressure, which increases the kinetic energy and thus the temperature of the moist air inside. This lowers its relative humidity, because warm air has a greater capacity to hold water *(recall Activity 14, step 9.)* The reverse occurs when this squeezing is relieved: less pressure reduces the kinetic energy and the temperature of the moist air, which *raises* its relative humidity to 100% (saturation). Water vapor now condenses into a visible cloud.

5. At lower elevations, breezes blowing off the sea are not fully saturated with water vapor. As this air rises, however, the resulting drop in pressure causes a drop in temperature, which reduces the air's capacity to hold water, and thus raises its relative humidity. When the humidity reaches 100% saturation, water begins to condense onto impurities in the air as tiny droplets, forming clouds or fog at higher elevations on the mountain slopes.

Materials

☐ A clear plastic, 2 liter soft-drink bottle with an air-tight lid. Colored plastics may reduce cloud visibility.
☐ Water.
☐ Matches.

(TO) construct an apparatus for use in the next activity that detects small changes in length. To geometrically calculate real movement based on apparent movement.

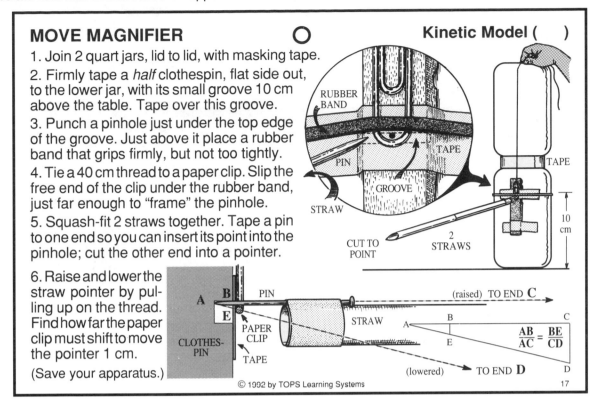

MOVE MAGNIFIER ○   Kinetic Model ( )

1. Join 2 quart jars, lid to lid, with masking tape.

2. Firmly tape a *half* clothespin, flat side out, to the lower jar, with its small groove 10 cm above the table. Tape over this groove.

3. Punch a pinhole just under the top edge of the groove. Just above it place a rubber band that grips firmly, but not too tightly.

4. Tie a 40 cm thread to a paper clip. Slip the free end of the clip under the rubber band, just far enough to "frame" the pinhole.

5. Squash-fit 2 straws together. Tape a pin to one end so you can insert its point into the pinhole; cut the other end into a pointer.

6. Raise and lower the straw pointer by pulling up on the thread. Find how far the paper clip must shift to move the pointer 1 cm.

(Save your apparatus.)

© 1992 by TOPS Learning Systems   17

Answers / Notes

6. BE = distance paper clip moves = unknown.
   CD = distance straw pointer moves = 1 cm
   AB = depth of clothespin notch + radius of paper clip wire = 3 mm *(see below)*
   AC = full length of pointer = 42.6 cm.

$$BE = \frac{AB}{AC} \times \frac{CD}{1} = \frac{3 \text{ mm}}{426 \text{ mm}} \times \frac{1 \text{ cm}}{1} = .007 \text{ cm} = .07 \text{ mm}$$

Students will find it difficult to accurately measure AB. Suggest they use a twist-tie: bare the wire at one end and insert it into the pinhole. Pull up on the paper clip, taking care to keep it pressed flush against the clothespin as you raise it. This will kink the wire where it meets the paper clip. Find the length of AB in mm by measuring from the end of the wire to the center of this kink.

Extension

Calibrate a scale at the end of your pointer that measures how far the paper clip actually moves. *(Based on results in the sample answer, centimeter lengths equally subdivided into sevenths would measure real paper clip movement in increments of .01 mm.)*

Materials

☐ Two quart jars with lids.
☐ Masking tape and scissors.
☐ A wooden clothespin half.
☐ A straight pin and rubber band.
☐ Thread.
☐ A paper clip.
☐ Two straws.
☐ The paper metric ruler.
☐ A twist-tie or other small diameter flexible wire.
☐ A calculator.

(TO) understand the expansion and contraction of aluminum foil in terms of the kinetic model. To measure its changing length.

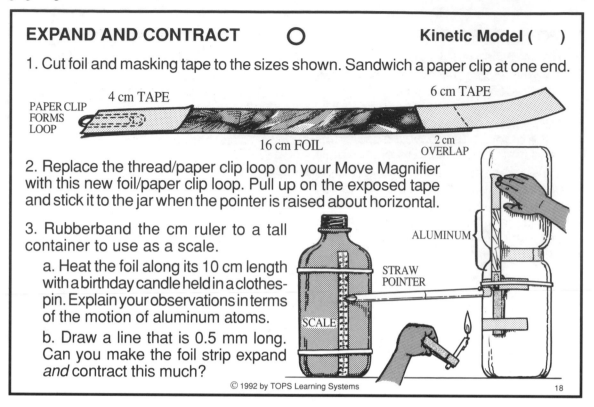

EXPAND AND CONTRACT ○ **Kinetic Model ( )**

1. Cut foil and masking tape to the sizes shown. Sandwich a paper clip at one end.

PAPER CLIP FORMS LOOP
4 cm TAPE
6 cm TAPE
16 cm FOIL
2 cm OVERLAP

2. Replace the thread/paper clip loop on your Move Magnifier with this new foil/paper clip loop. Pull up on the exposed tape and stick it to the jar when the pointer is raised about horizontal.

3. Rubberband the cm ruler to a tall container to use as a scale.

ALUMINUM

STRAW POINTER

SCALE

    a. Heat the foil along its 10 cm length with a birthday candle held in a clothespin. Explain your observations in terms of the motion of aluminum atoms.

    b. Draw a line that is 0.5 mm long. Can you make the foil strip expand *and* contract this much?

© 1992 by TOPS Learning Systems
18

Answers / Notes

3. *Those students who have calibrated their own scales (see the enrichment section of the previous activity) should use them here. They will be able to answer question 3b simply by reading their scale.*

3a. The pointer falls as the foil heats up, showing that it expands. This is predicted by the kinetic theory since heat increases the motion of the aluminum atoms within the strip, causing them to jostle and bump each other more vigorously and thus take up a little more room. This process reverses as the foil cools down. The pointer now rises as the aluminum contracts. Vibrating less vigorously, the aluminum atoms take up less space.

3b. 0.5 mm line:

In the previous activity, your students used similar triangles to find how far the paper clip moved in relation to the pointer. This proportion should be applied here. In our particular case:

$$\frac{0.07 \text{ mm}}{1 \text{ cm}} = \frac{0.5 \text{ mm}}{x} \; ; \quad x = \frac{0.5 \text{ mm}}{0.07 \text{ mm}} \approx 7 \text{ cm}$$

If the candle is held very near, so its flame touches the 10 cm of exposed foil along its entire surface, the pointer drops, then rises, 7 cm and more when you remove the flame. *(The pointer may even drop off the scale and touch the table surface, especially if a hotter flame source is used. If more height is needed, rest the jars on a can or other object.)*

Materials

☐ Aluminum foil.
☐ Masking tape.
☐ Scissors.
☐ A paper clip.
☐ The paper cm ruler.
☐ The Move Magnifier apparatus constructed in the previous activity.

☐ Any container, round or rectangular, that is at least 20 cm tall. The 2 liter bottle from activity 16 is suitable.
☐ Rubber bands.
☐ A birthday candle and clothespin.
☐ An empty can (if needed) to elevate the Move Magnifier relative to the scale.
☐ A calculator.

(TO) observe how a bi-material strip bends as its dissimilar sides expand and contract by different amounts. To recognize its application in a thermostat.

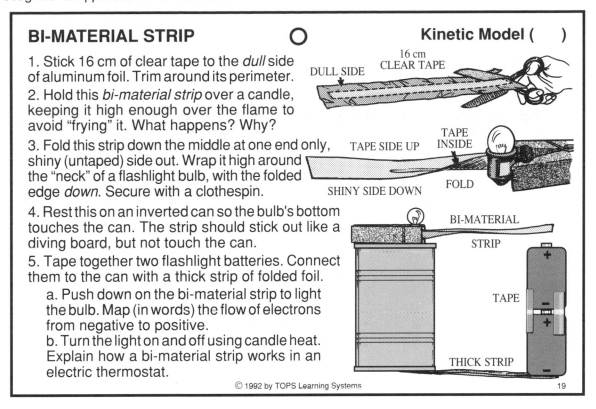

BI-MATERIAL STRIP ⭕ Kinetic Model ( )

1. Stick 16 cm of clear tape to the *dull* side of aluminum foil. Trim around its perimeter.

2. Hold this *bi-material strip* over a candle, keeping it high enough over the flame to avoid "frying" it. What happens? Why?

3. Fold this strip down the middle at one end only, shiny (untaped) side out. Wrap it high around the "neck" of a flashlight bulb, with the folded edge *down*. Secure with a clothespin.

4. Rest this on an inverted can so the bulb's bottom touches the can. The strip should stick out like a diving board, but not touch the can.

5. Tape together two flashlight batteries. Connect them to the can with a thick strip of folded foil.

   a. Push down on the bi-material strip to light the bulb. Map (in words) the flow of electrons from negative to positive.

   b. Turn the light on and off using candle heat. Explain how a bi-material strip works in an electric thermostat.

19

Answers / Notes

2. The bi-material strip bends toward its aluminum side as it heats up; away from its aluminum side as it cools down. On heating, the tape expands faster, bending towards the slightly shorter aluminum. On cooling, the tape contracts faster, bending away from the slightly longer aluminum.

3. *Details in this step are critically important if the bulb is to light in step 5: if you fold the dull (taped) side out, only the tape will contact the bulb. If you position the folded edge aiming up, the bi-material strip will bend the wrong way and will not complete the circuit.*

4. *If the bottom of the bi-material "diving board" touches the can, it will short out the bulb when the circuit is completed in step 5.*

5. *A single layer of foil is too thin to guarantee electrical contact between the bottom of the batteries and the bottom rim of the can. Multiple folds are needed.*

5a. From the negative (bottom) terminal, electrons flow through the foil strip, up the can, through the bulb, through the bi-material strip, into the positive (top) terminal.

5b. *To turn the light on and off with candle heat, position the free end of the bi-material strip over the top battery terminal. Heating it from below lengthens the tape faster than the foil, bending it downward to establish contact. On cooling, the strip bends back up and breaks contact.*

A thermostat is controlled by a bi-material strip that works like a thermal switch. *(These are normally called bimetal strips because they are composed of two metals.)* On cooling, it bends towards an electrical contact, completes the circuit and turns on the heat. On warming, it bends away from this electric contact, breaks the circuit and turns the heat back off. *(Notice that this works just opposite to the hot/on, cool/off switch modeled above.)*

Materials

☐ Clear tape.
☐ The 20 cm ruler.
☐ Aluminum foil.
☐ Scissors.

☐ A birthday candle.
☐ Matches.
☐ A flashlight bulb.
☐ A clothespin.

☐ A tin can. It should be about as tall as two batteries end to end. Vegetable cans are about the right size.
☐ Two size-D flashlight batteries.
☐ Masking tape.

(TO) calculate the heat of fusion for ice.

HEAT OF FUSION ⃝ Kinetic Model (   )

1. Start with 100 ml of warm (30° C) water in a graduated cylinder, plus an ice cube that is just beginning to melt.

   a. Pour the warm water into a styrofoam cup, then take its **initial temperature**.

   b. Pat the ice cube dry with a paper towel, then put it in the water. Stir constantly.

   c. Record the **lowest temperature** reached by the water and its **final volume**.

2. You know that 1 ml of water has a mass of 1 g; that each change of 1° C in this much water is a gain or loss of 1 calorie of heat. Calculate the *heat of fusion,* the cal/g needed to melt ice. (Hint: Find A,B,C,D in order. $D + C = A$.)

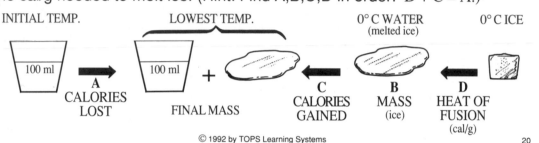

| INITIAL TEMP. | LOWEST TEMP. | 0° C WATER (melted ice) | 0° C ICE |

A CALORIES LOST   FINAL MASS   **C** CALORIES GAINED   **B** MASS (ice)   **D** HEAT OF FUSION (cal/g)

20

Answers / Notes

1-2. Sample answer:

A:

    initial temperature of warm water = 29.5° C
    –lowest temperature of mix = 15.8° C
    _____
    temperature loss = 13.7° C
    calories lost by warm water = (13.7° C) (100 g) = 1370 cal

B:

    final amount of water = 115 ml = 115 g
    – original amount of water = 100 ml = 100 g
    _____
    original amount of ice =   15 ml =   15 g

C:

    final temperature of mix = 15.8° C
    initial temperature of cold water =   0.0° C
    _____
    temperature gain = 15.8° C
    calories gained by cold water = (15.8° C) (15g) = 237 cal

D:

    $D + C = A$
    heat of fusion + 237 cal = 1370 cal
    heat of fusion = 1133 cal for 15 g ice
    heat of fusion = 76 cal/g
    (accepted value = 80 cal/g)

Materials

☐ Warm tap water.
☐ A 100 ml graduate.

☐ A thermometer.
☐ An ice cube from a regular-sized ice tray.

☐ A paper towel.
☐ A styrofoam cup.

(TO) quantitatively measure the volume of air inside a test tube at two different temperatures.

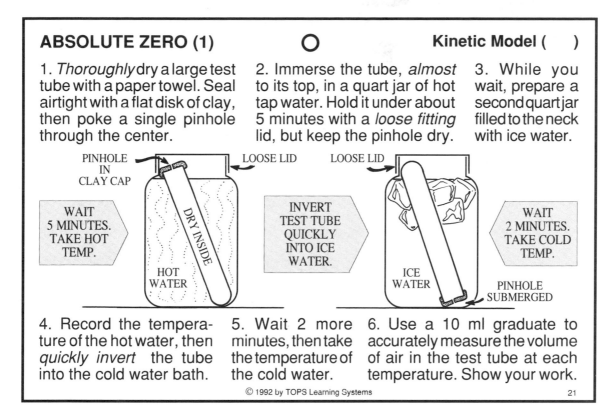

ABSOLUTE ZERO (1)   O   Kinetic Model (   )

1. *Thoroughly* dry a large test tube with a paper towel. Seal airtight with a flat disk of clay, then poke a single pinhole through the center.

2. Immerse the tube, *almost* to its top, in a quart jar of hot tap water. Hold it under about 5 minutes with a *loose fitting* lid, but keep the pinhole dry.

3. While you wait, prepare a second quart jar filled to the neck with ice water.

PINHOLE IN CLAY CAP   LOOSE LID   LOOSE LID

WAIT 5 MINUTES. TAKE HOT TEMP.

DRY INSIDE

HOT WATER

INVERT TEST TUBE QUICKLY INTO ICE WATER.

ICE WATER

PINHOLE SUBMERGED

WAIT 2 MINUTES. TAKE COLD TEMP.

4. Record the temperature of the hot water, then *quickly invert* the tube into the cold water bath.

5. Wait 2 more minutes, then take the temperature of the cold water.

6. Use a 10 ml graduate to accurately measure the volume of air in the test tube at each temperature. Show your work.

© 1992 by TOPS Learning Systems   21

Answers / Notes

2. *The pinhole must remain above water, and the jar lid must be loose enough to vent moist air. Otherwise the dry air inside the test tube may become humidified and skew the results.*

4-6. temperature of warm air = temperature of hot bath = 46° C.
    volume of warm air = volume of test tube = 28.2 ml.
    temperature of cold air = temperature of cold bath = 5° C.
    volume of cold air = volume of test tube − volume of water intake = 28.2 ml − 3.7 ml = 24.5 ml

If necessary, review how to measure liquid volumes with a graduated cylinder. Note that most 10 ml graduates are calibrated in 0.2 ml increments, not 0.1 ml increments.

Materials

☐ A large test tube. For greatest experimental accuracy, use your largest test tubes that still fit inside quart jars. We used a 15 cm tube with 16 mm inside diameter.
☐ A paper towel.
☐ Oil-based modeling clay.
☐ A straight pin.
☐ Two quart jars with at least one lid.
☐ Hot water from a tap or some other heat source. Use the hottest possible water that can still be handled safely and won't crack your jars.
☐ A wall clock.
☐ A thermometer.
☐ Ice water. Supply plenty of ice cubes to cool the water near 0° C.
☐ A 10 ml graduated cylinder. This is essential to accurately measure the small quantity of water that enters the cooling test tube. The test tube volume can be measured with this small graduate as well, or more conveniently, with a 100 ml graduate.

(TO) experimentally calculate absolute zero and compare it with the accepted value. To appreciate the physical significance of this ultimate cold.

ABSOLUTE ZERO (2)　　　○　　　　**Kinetic Model (　)**

1. Cut out the 2 Absolute Zero half-graphs along the dotted lines. Align the grids along the vertical axis, then tape together.

2. Accurately plot and circle 2 data points that summarize your results from the previous activity.

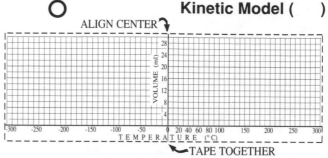

3. As air cools (under constant pressure), its volume shrinks in *direct proportion* to its drop in temperature. The average kinetic energy of its molecules decreases until they contract into liquid air with near-zero volume. This temperature where all thermal motion stops is called *absolute zero*.

    a. Extrapolate your two data points: extend a straight line back to zero volume and read the temperature at that point.

    b. Mark −273° C on your zero volume line. How close did you get to this accepted value for absolute zero? What limits your accuracy in this experiment?

4. Model absolute zero using a plastic bag and corn seeds as a kinetic model.

    a. Explain how you did this.

    b. Can air get colder than −273° C? Explain.

　　　22

Answers / Notes

2-3.

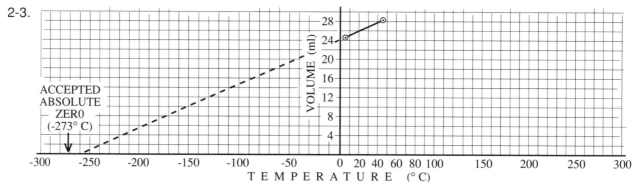

Our particular value for absolute zero crossed the zero volume axis at about −260° C. This is 13° C warmer than the accepted value. Accuracy is limited in this experiment by extrapolating over such a long distance relative to the spacing between data points. Small experimental errors produce large extrapolation errors.

4a. Place the corn seeds in the plastic bag puffed up with air, in the usual manner. This time don't shake them at all; let them lay in a little pile at the bottom. At absolute zero a gas has no thermal motion.

4b. Nothing can get colder than −273° C. Molecules move slower and slower in response to cooling temperatures until they finally stop. They can move no slower. Air that shrinks to zero volume can shrink no more.

Materials

☐ The 2 Absolute Zero half-graphs. Photocopy these from the supplementary page.
☐ Scissors.
☐ Clear tape.
☐ Temperature and volume data from the previous activity.
☐ A long straightedge. Folded scratch paper will serve.
☐ A plastic produce bag with popcorn seeds.

(TO) use the absolute zero graph to plot the average temperature of air inside a heated test tube.

ABSOLUTE ZERO (3)   ○   Kinetic Model (   )

1. Set out a bowl and a jar nearly filled with warm water.
2. Clamp a clothespin handle just below the mouth of a test tube.

3. Drive off any moisture by heating it *gently* in an upright position. (CAUTION: Hot glass can burn you severely. Don't let its cool appearance fool you!)

4. Trap the hot dry air in the tube by inverting it at an angle. Continue heating for 1 minute, always keeping the mouth *below* the flame.

5. *Quickly* plunge the upended test tube by its handle into the bowl of water. Pour the jar of water over it until completely cool. Then trap *all* the water that enters with your thumb, and turn it right-side-up.

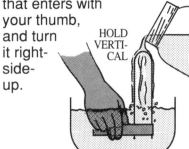

HOLD VERTI-CAL

6. Use a graduated cylinder, a thermometer and your graph to find how hot you heated the air in the tube. *(Hint: use Absolute Zero and one experimental point.)*

© 1992 by TOPS Learning Systems   23

Answers / Notes

3-5. CAUTION: Your students will be handling extremely hot glass. Close supervision is absolutely essential.

6. Here are the results for a small 15.4 ml test tube heated in a candle flame:

volume of heated air = volume of test tube = 15.4 ml.

volume of cooled air = volume of test tube – volume of water intake = 15.4 ml – 6.7 ml = 8.7 ml

temperature of air after cooling = temperature of water bath = 28.5° C.

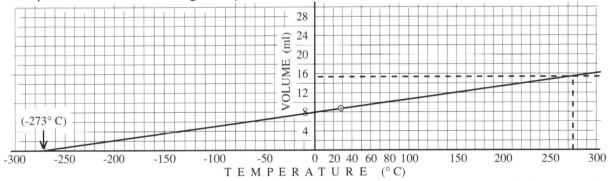

Two points on the graph define a directly proportional straight line: absolute zero at (-273,0) plus the final temperature and volume of the cooled air at (28.5 , 8.7). Extending this line to intersect 15.4 ml (the total volume of heated air) defines a temperature of 275° C at that volume.

Materials

☐ A bowl of warm tap water (cold water may crack the test tube).
☐ A jar of warm tap water.
☐ A test tube, large or small. Since it will be stressed by wide temperature variations, use a size that you don't mind losing, should it crack on cooling.

☐ A clothespin.
☐ A candle and matches.
☐ A wall clock or wristwatch.
☐ A long straightedge.
☐ A graduated cylinder and a thermometer.
☐ The absolute-zero graph from the previous activity.

(TO) boil warm water in a syringe at low pressure. To interpret a simple phase diagram.

A COOL BOIL ⭕ Kinetic Model (   )

1. In Activity 7 you boiled water.
   a. What boiling temperature did you record?
   b. According to this graph, what was your atmospheric pressure during that experiment?
   c. How would your results change at the top of Mt. Everest? On the shores of the Dead Sea?

2. Add 40° C water to a styrofoam cup.
   a. Is this water comfortable to touch? Can it boil at this temperature?
   b. Boil this luke-warm water inside a 3 cc syringe! Explain how you did it.

3. Water exists as a liquid *and* a gas only on the boiling point curve. Off this curve it is one *or* the other. Describe its phase(s) at:
   a. 40° C, 1 atm.   d. 90° C, 1 atm.
   b. 40° C, 0.08 atm.   e. 100° C, 1 atm.
   c. 40° C, 0.04 atm.   f. 110° C, 1 atm.

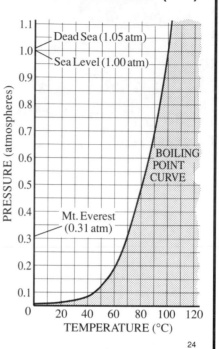

© 1992 by TOPS Learning Systems

24

Answers / Notes

1a. 100° C or slightly lower.

1b. 1 atmosphere or slightly lower.

1c. At the top of Mt. Everest, under less than 1/3 of an atmosphere, water boils at a cool 70° C. On the shores of the Dead Sea, under slightly more than 1 atmosphere pressure, it boils at a slightly hotter 101° C.

   Point out that the centigrade temperature scale is defined by the properties of water under a sea level pressure of 1 atmosphere: boiling is arbitrarily set at "100"; freezing at "0." These figures are definitions, not measurements.

2a. Water at 40° C is comfortably warm. It boils at this temperature if the pressure is reduced below about 0.08 atmosphere.

2b. Draw in enough 40° C water to fill the syringe perhaps 1/6 full. Close the intake with your finger so no air can enter, then pull back on the plunger to reduce the inside pressure below 0.08 atm. The water bubbles in a cool boil.

3a. liquid   3d. liquid
3b. liquid and gas 3e. liquid and gas
3c. gas   3f. gas

The y-axis of this graph essentially defines a freezing boundary between water and ice. Of special interest is the triple point—the temperature (near 0° C) and pressure (near 0 atm) where water coexists in all three states: amazing freezing bubbles of boiling water!

Materials

☐ Experimental results from activity 7.
☐ Warm water from a tap.
☐ A styrofoam cup.
☐ A thermometer.
☐ A clear plastic 3 cc syringe.

REPRODUCIBLE
STUDENT
TASK CARDS

Task Cards Options

Here are 3 management options to consider before you photocopy:

1. Consumable Worksheets: Copy 1 complete set of task card pages. Cut out each card and fix it to a separate sheet of boldly lined paper. Duplicate a class set of each worksheet master you have made, 1 per student. Direct students to follow the task card instructions at the top of each page, then respond to questions in the lined space underneath.

2. Nonconsumable Reference Booklets: Copy and collate the 2-up task card pages in sequence. Make perhaps half as many sets as the students who will use them. Staple each set in the upper left corner, both front and back to prevent the outside pages from working loose. Tell students that these task card booklets are for reference only. They should use them as they would any textbook, responding to questions on their own papers, returning them unmarked and in good shape at the end of the module.

3. Nonconsumable Task Cards: Copy several sets of task card pages. Laminate them, if you wish, for extra durability, then cut out each card to display in your room. You might pin cards to bulletin boards; or punch out the holes and hang them from wall hooks (you can fashion hooks from paper clips and tape these to the wall); or fix cards to cereal boxes with paper fasteners, 4 to a box; or keep cards on designated reference tables. The important thing is to provide enough task card reference points about your classroom to avoid a jam of too many students at any one location. Two or 3 task card sets should accommodate everyone, since different students will use different cards at different times.

INDIRECT EVIDENCE O   Kinetic Model (   )

1. Tape the square, circle and triangle patterns to cardboard. Carefully cut out each shape.

2. Put one shape into a clean, dry milk carton. Invent ways to identify it *without* touching it or looking at it directly.

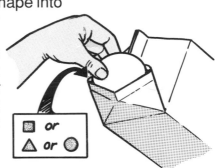

3. Refine your methods until you can correctly identify each shape that a friend hides inside. Explain your technique.

4. Scientists tell us that air is made from tiny *molecules* that are too small to see. How can they know about things they haven't seen?

1

MODELING UNKNOWNS O   Kinetic Model (   )

1. Get a sealed "mystery box." Experiment to determine the properties of the object inside, without unsealing your box.

   a. Sketch the size and shape of each object inside.

   b. Cite experimental evidence to support your drawing.

   c. Compare the properties of each unknown object to known objects that you put inside a second milk carton.

2. Name a physical property of your object that you can't be sure about.

3. Repeat these steps with other mystery boxes coded with other letters.

4. To correctly model real-world science, is it fair to remove the tape from any mystery box and peek inside? Defend your answer.

2

A KINETIC MODEL   O   Kinetic Model (   )

1. Fully open a plastic bag, then twist its mouth closed to form a balloon. What does this "mystery bag" contain? Describe its contents as fully as possible.

2. Set a jar of water next to a jar of air. If you magnified the molecules in each about twenty million times, they would have these relative sizes:

a. Identify the main elements in water and air molecules.

b. These molecules are nearly the same size, yet water is plainly visible, and air is not. Propose an explanation.

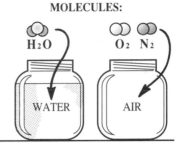

MOLECULES:

H_2O O_2 N_2

WATER   AIR

3. Let kernels of popcorn model air molecules. Vigorously shake a handful of these "molecules" inside the plastic balloon from step 1.

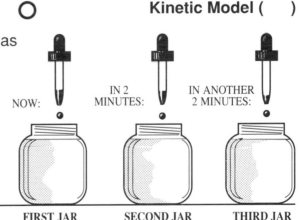

a. What properties of air does this model help to explain?

b. Why call this popcorn representation a *kinetic* model?

c. Does this kinetic model suggest that all air molecules move at the same speed? Observe closely.

d. Does this model fail to act like an actual bag of air in any respect?

© 1992 by TOPS Learning Systems   3

KINETIC MIX?   O   Kinetic Model (   )

1. Fill 3 small jars with water that has come to room temperature. Set them on a solid surface, free of vibrations, so the water becomes very calm.

2. Add 1 drop of food coloring to the first jar. At 2 minute intervals, repeat for the second and third jars. Each drop should cleanly separate from the dropper *before* it touches the water.

NOW: IN 2 MINUTES: IN ANOTHER 2 MINUTES:

FIRST JAR   SECOND JAR   THIRD JAR

a. Describe the color patterns that form in the water. How do they change over time?

b. Do you think the kinetic model for air applies to water? Why?

c. How might these jars look after 24 hours? Make a prediction.

3. Test your prediction with a fresh jar of calm water resting in a safe, out-of-the-way place.

© 1992 by TOPS Learning Systems   4

AIR THERMOMETER Kinetic Model (   )

1. Connect narrow tubing to a large test tube, sealing it with a clay cap or one-hole stopper.

2. Uncap the test tube. Place several drops of colored water at the center of the tubing to form a plug, then reseal.

3. Dip the test tube first in warm water, then in cool water.
  a. How does the water plug respond? Propose a theory.
  b. Compare and contrast this device with a lab thermometer.

COLORED WATER PLUG

AIR

WARM WATER

COOL WATER

4. Energy of motion is called *kinetic energy*. How do scientists measure the kinetic energy of atoms and molecules?

5

PHASE CHANGES Kinetic Model (   )

1. Water exists in 3 *phases* — solid (ice), liquid (water) and gas (vapor). Fully describe each *phase change* below using the capitalized vocabulary.

  a. CONDENSE: Tape a strip of foil around a small jar, then drop 2 ice cubes inside. Observe the foil.
  b. FREEZE: Add just enough salt to bury the ice, then seal with a lid. Swirl this around the wall of the jar for several minutes, holding it top and bottom with a paper towel.
  c. MELT: Remove the frosted foil from the jar, while keeping it under close observation.
  d. EVAPORATE: Lightly warm the water droplets on the foil over a flame.

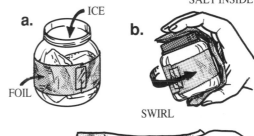

a. ICE

FOIL

b. SALT INSIDE

SWIRL

c. REMOVE FOIL

d. HEAT GENTLY

2. During which phase changes do water molecules move faster? Slower?

3. During which phase changes, therefore, must water molecules gain heat? Release heat?

6

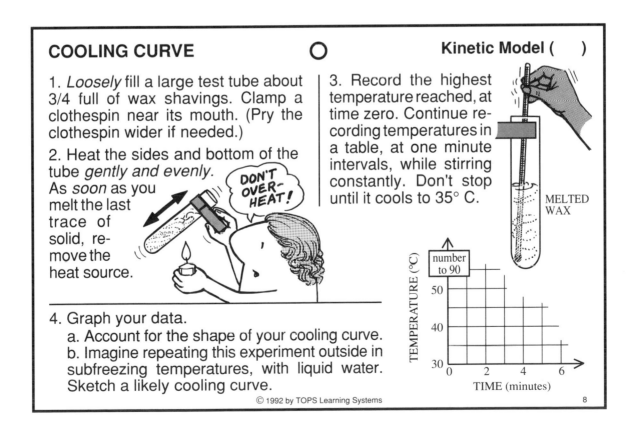

HEATING CURVE ○   **Kinetic Model (   )**

1. Clamp a clothespin handle near the mouth of a large test tube. Pry it wider, as necessary, to fit around the tube.

2. Fill the tube with snow or ice shavings. Insert a thermometer to the bottom to record the lowest temperature. Call this time zero.

3. Immediately heat the ice over a candle flame while stirring with a thermometer. Ask a friend to record the temperature every 30 seconds in a data table. Continue until the water has boiled for 2 minutes.

4. Plot your data on a graph. Circle each point, then connect them with a smooth, curving line.

5. Does the temperature of water always rise as water absorbs heat? Explain.

TIME (min)	TEMP. (°C)
0	
0.5	
1.0	
1.5	
:	
:	

number to 100

TEMPERATURE (°C)
20
10
0

TIME (minutes)
1 2

© 1992 by TOPS Learning Systems   7

COOLING CURVE ○   **Kinetic Model (   )**

1. *Loosely* fill a large test tube about 3/4 full of wax shavings. Clamp a clothespin near its mouth. (Pry the clothespin wider if needed.)

2. Heat the sides and bottom of the tube *gently and evenly*. As *soon* as you melt the last trace of solid, remove the heat source.

DON'T OVER- HEAT!

3. Record the highest temperature reached, at time zero. Continue recording temperatures in a table, at one minute intervals, while stirring constantly. Don't stop until it cools to 35° C.

MELTED WAX

4. Graph your data.
   a. Account for the shape of your cooling curve.
   b. Imagine repeating this experiment outside in subfreezing temperatures, with liquid water. Sketch a likely cooling curve.

number to 90

TEMPERATURE (°C)
50
40
30

TIME (minutes)
0 2 4 6

© 1992 by TOPS Learning Systems   8

BLOW-UP   O   **Kinetic Model ( )**

1. Cut a piece of aluminum foil about as big as an index card. Fold to 1/4 size, forming an open pocket.

2. Deposit just *one* drop of water into a corner. Seal both open edges with 2 tight folds each.

3. Heat the moist corner over a flame, after clamping the opposite corner in a clothespin. Use the kinetic model to interpret your observations.

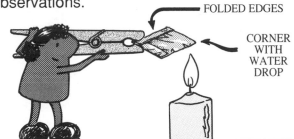

FOLDED EDGES

CORNER WITH WATER DROP

4. Cut more foil to the size of an index card. Shape it around a large test tube to fashion a "mini-pan" with a long handle.

    a. Heat a single popcorn over a flame for at least a minute, shaking it *constantly*. What makes the popcorn pop?

    b. What is the essential difference between liquid water and water vapor?

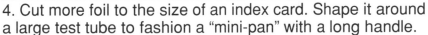

TEST TUBE MOLD

MINI-PAN

© 1992 by TOPS Learning Systems   9

COOL IT!   O   **Kinetic Model ( )**

1. Line 2 cups with plastic sandwich bags, then fill each half full with room temperature water.

2. Twist and paper-clip one bag closed; leave the other open. Over a long period of time, would you expect to observe changes in either bag? Explain.

3. Push the nose of a thermometer deep into the side of each bag, taking the temperature of its water *through* the plastic. Record your results to the nearest 0.1° C.

CLOSED BAG:

OPEN BAG:

ROOM TEMPERATURE WATER

4. As water evaporates, which molecules are more likely to move into the gas phase — slow moving molecules or fast moving ones? Why?

5. Reconcile your answer in step 4 with your experimental result in step 3.

© 1992 by TOPS Learning Systems   10

INCREDIBLE JOURNEY ○ Kinetic Model (   )

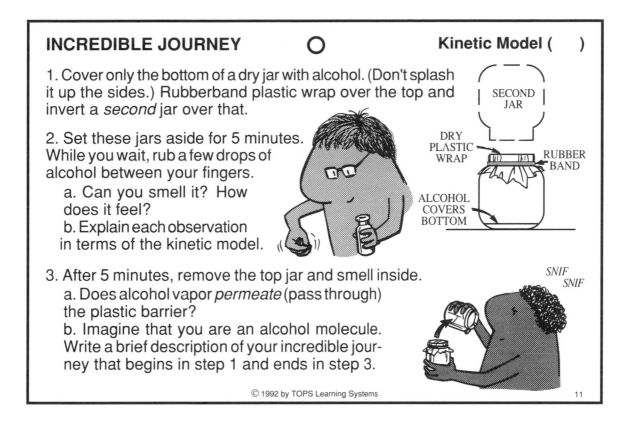

1. Cover only the bottom of a dry jar with alcohol. (Don't splash it up the sides.) Rubberband plastic wrap over the top and invert a *second* jar over that.

2. Set these jars aside for 5 minutes. While you wait, rub a few drops of alcohol between your fingers.
    a. Can you smell it? How does it feel?
    b. Explain each observation in terms of the kinetic model.

3. After 5 minutes, remove the top jar and smell inside.
    a. Does alcohol vapor *permeate* (pass through) the plastic barrier?
    b. Imagine that you are an alcohol molecule. Write a brief description of your incredible journey that begins in step 1 and ends in step 3.

© 1992 by TOPS Learning Systems   11

TIPPING THE BALANCE ○ Kinetic Model (   )

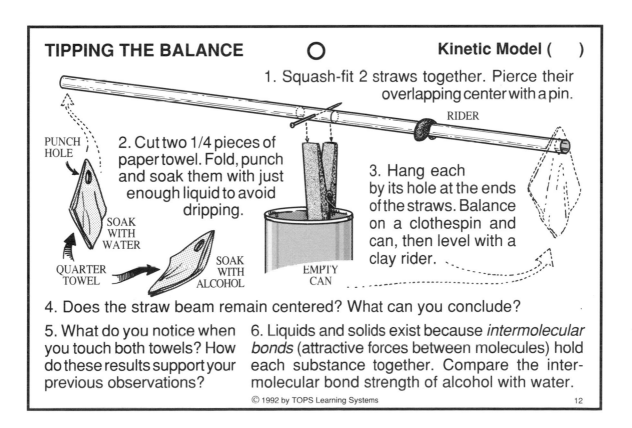

1. Squash-fit 2 straws together. Pierce their overlapping center with a pin.

2. Cut two 1/4 pieces of paper towel. Fold, punch and soak them with just enough liquid to avoid dripping.

3. Hang each by its hole at the ends of the straws. Balance on a clothespin and can, then level with a clay rider.

4. Does the straw beam remain centered? What can you conclude?

5. What do you notice when you touch both towels? How do these results support your previous observations?

6. Liquids and solids exist because *intermolecular bonds* (attractive forces between molecules) hold each substance together. Compare the intermolecular bond strength of alcohol with water.

© 1992 by TOPS Learning Systems   12

RELATIVE HUMIDITY ○ Kinetic Model (   )

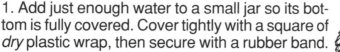

1. Add just enough water to a small jar so its bottom is fully covered. Cover tightly with a square of *dry* plastic wrap, then secure with a rubber band.

2. Deposit a drop of water on the left half of the plastic, then quickly put a drop of alcohol on the right.

3. Look at each drop with a magnifying lens. What begins to happen underneath? How fast?

4. Fan both drops about 200 times with a folded sheet of paper, then wipe them dry with a tissue.

    a. Why are water vapor molecules more likely to condense under the drops than elsewhere on the plastic wrap? Why under one specific drop?

    b. Does the area under each drop show *uniform* cooling? Why?

5. *Relative humidity* measures the capacity of air to hold water vapor. If the jar air has a relative humidity of 100% this means no additional water can evaporate into it without an equal amount also condensing back out.

    a. Why do your droplets remain the same size?

    b. How could you make them disappear?

WATER DROPS

EVAP. | COND.

WATER VAPOR

100% HUMIDITY
(SATURATED AIR)

13

HYGROMETER ○ Kinetic Model (   )

1. Rubber-band 2 thermometers to a clothespin so they hang over the rim of a quart jar.

2. Tie a tissue paper "skirt" onto the outside bulb with string. Adjust this skirt to just touch the bottom of an upturned lid.

3. Connect jar and lid by setting both on a strip of tape.

4. Cut out a *relative humidity table* and tape it to the jar.

5. Gently rotate the thermometers so you can read both from the front. Record the current air temperature on both instruments.

HANG ONE THERMO-METER INSIDE JAR

SKIRT

LID

WATER

TAPE LID TO JAR

6. Fill the lid with water. Only the tissue skirt should hang in the water. Do both thermometers still read the same? Why?

7. Use this *hygrometer* to find your current relative humidity. Tell how you did this.

8. Cover the bottom of the quart jar with water, then rest the hygrometer over a pencil so *both* thermometers hang *inside*. Does the air in the jar have a higher relative humidity? Why?

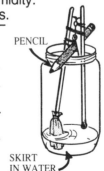

PENCIL

SKIRT IN WATER

9. How would you expect the humidity to change if you held your hygrometer over a radiator or hot plate at 30° C? Test your prediction, if possible.

14

KINETIC PUSH-UPS ○   Kinetic Model (   )

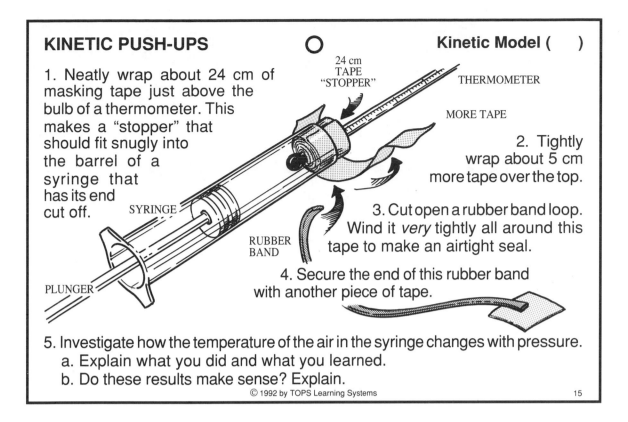

1. Neatly wrap about 24 cm of masking tape just above the bulb of a thermometer. This makes a "stopper" that should fit snugly into the barrel of a syringe that has its end cut off.

SYRINGE

RUBBER BAND

PLUNGER

24 cm TAPE "STOPPER"

THERMOMETER

MORE TAPE

2. Tightly wrap about 5 cm more tape over the top.

3. Cut open a rubber band loop. Wind it *very* tightly all around this tape to make an airtight seal.

4. Secure the end of this rubber band with another piece of tape.

5. Investigate how the temperature of the air in the syringe changes with pressure.
   a. Explain what you did and what you learned.
   b. Do these results make sense? Explain.

© 1992 by TOPS Learning Systems   15

CLOUDS ○   Kinetic Model (   )

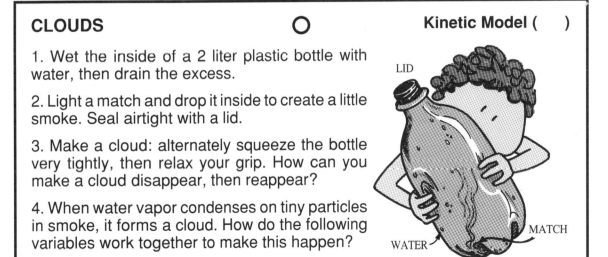

1. Wet the inside of a 2 liter plastic bottle with water, then drain the excess.

2. Light a match and drop it inside to create a little smoke. Seal airtight with a lid.

3. Make a cloud: alternately squeeze the bottle very tightly, then relax your grip. How can you make a cloud disappear, then reappear?

4. When water vapor condenses on tiny particles in smoke, it forms a cloud. How do the following variables work together to make this happen?

LID

WATER

MATCH

SQUEEZE AND RELEASE

   ☞ **PRESSURE**
   ☞ **TEMPERATURE**
   ☞ **RELATIVE HUMIDITY**

5. A steady breeze blows off the sea and over a range of mountains. Clouds gather at higher elevations but not below. What is happening?

© 1992 by TOPS Learning Systems   16

MOVE MAGNIFIER   ○   Kinetic Model ( )

1. Join 2 quart jars, lid to lid, with masking tape.

2. Firmly tape a *half* clothespin, flat side out, to the lower jar, with its small groove 10 cm above the table. Tape over this groove.

3. Punch a pinhole just under the top edge of the groove. Just above it place a rubber band that grips firmly, but not too tightly.

4. Tie a 40 cm thread to a paper clip. Slip the free end of the clip under the rubber band, just far enough to "frame" the pinhole.

5. Squash-fit 2 straws together. Tape a pin to one end so you can insert its point into the pinhole; cut the other end into a pointer.

6. Raise and lower the straw pointer by pulling up on the thread. Find how far the paper clip must shift to move the pointer 1 cm.

(Save your apparatus.)

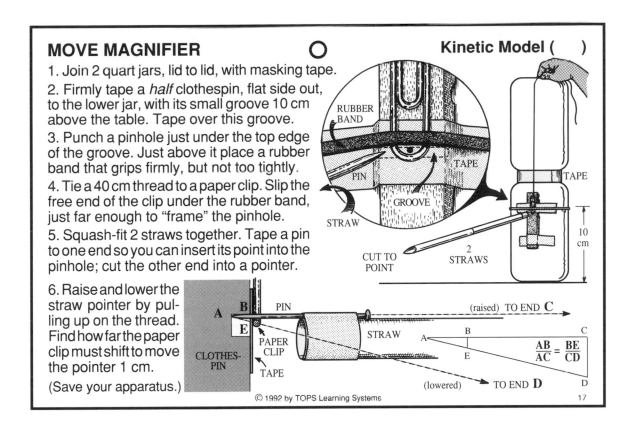

$$\frac{AB}{AC} = \frac{BE}{CD}$$

© 1992 by TOPS Learning Systems   17

EXPAND AND CONTRACT   ○   Kinetic Model ( )

1. Cut foil and masking tape to the sizes shown. Sandwich a paper clip at one end.

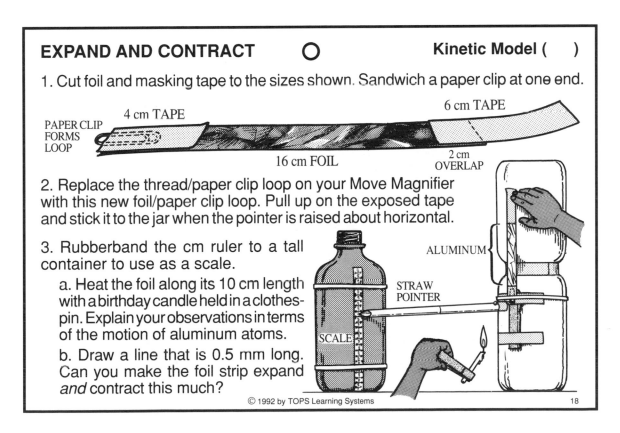

PAPER CLIP FORMS LOOP

4 cm TAPE

16 cm FOIL

6 cm TAPE

2 cm OVERLAP

2. Replace the thread/paper clip loop on your Move Magnifier with this new foil/paper clip loop. Pull up on the exposed tape and stick it to the jar when the pointer is raised about horizontal.

3. Rubberband the cm ruler to a tall container to use as a scale.

ALUMINUM

STRAW POINTER

SCALE

   a. Heat the foil along its 10 cm length with a birthday candle held in a clothespin. Explain your observations in terms of the motion of aluminum atoms.

   b. Draw a line that is 0.5 mm long. Can you make the foil strip expand *and* contract this much?

© 1992 by TOPS Learning Systems   18

BI-MATERIAL STRIP   ◯   Kinetic Model (   )

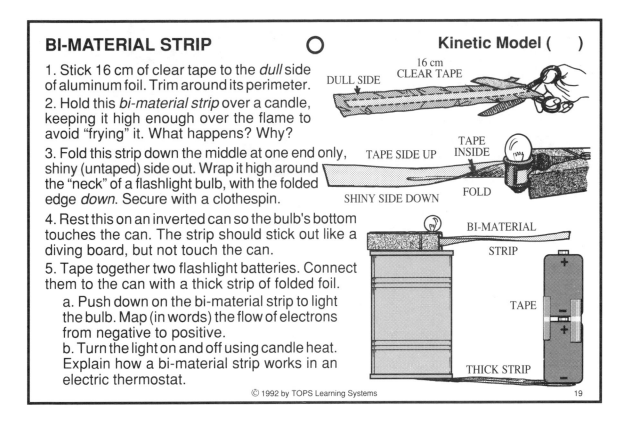

1. Stick 16 cm of clear tape to the *dull* side of aluminum foil. Trim around its perimeter.

2. Hold this *bi-material strip* over a candle, keeping it high enough over the flame to avoid "frying" it. What happens? Why?

3. Fold this strip down the middle at one end only, shiny (untaped) side out. Wrap it high around the "neck" of a flashlight bulb, with the folded edge *down*. Secure with a clothespin.

4. Rest this on an inverted can so the bulb's bottom touches the can. The strip should stick out like a diving board, but not touch the can.

5. Tape together two flashlight batteries. Connect them to the can with a thick strip of folded foil.

    a. Push down on the bi-material strip to light the bulb. Map (in words) the flow of electrons from negative to positive.

    b. Turn the light on and off using candle heat. Explain how a bi-material strip works in an electric thermostat.

© 1992 by TOPS Learning Systems   19

HEAT OF FUSION   ◯   Kinetic Model (   )

1. Start with 100 ml of warm (30° C) water in a graduated cylinder, plus an ice cube that is just beginning to melt.

    a. Pour the warm water into a styrofoam cup, then take its **initial temperature**.

    b. Pat the ice cube dry with a paper towel, then put it in the water. Stir constantly.

    c. Record the **lowest temperature** reached by the water and its **final volume**.

2. You know that 1 ml of water has a mass of 1 g; that each change of 1° C in this much water is a gain or loss of 1 calorie of heat. Calculate the *heat of fusion,* the cal/g needed to melt ice. (Hint: Find A,B,C,D in order. D + C = A.)

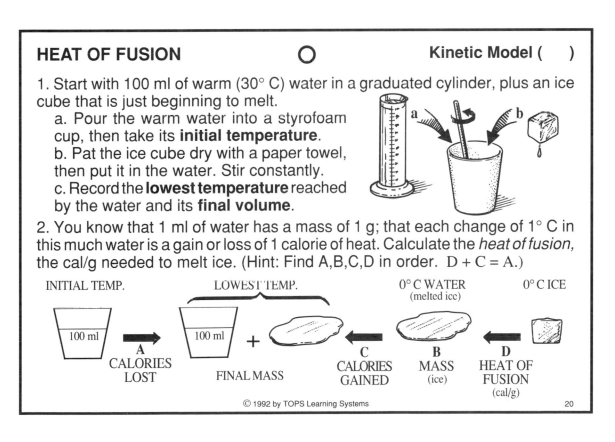

© 1992 by TOPS Learning Systems   20

ABSOLUTE ZERO (1) ◯ Kinetic Model (   )

1. *Thoroughly* dry a large test tube with a paper towel. Seal airtight with a flat disk of clay, then poke a single pinhole through the center.

2. Immerse the tube, *almost* to its top, in a quart jar of hot tap water. Hold it under about 5 minutes with a *loose fitting* lid, but keep the pinhole dry.

3. While you wait, prepare a second quart jar filled to the neck with ice water.

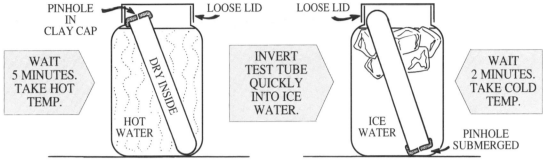

PINHOLE IN CLAY CAP — LOOSE LID

LOOSE LID

WAIT 5 MINUTES. TAKE HOT TEMP.

DRY INSIDE

HOT WATER

INVERT TEST TUBE QUICKLY INTO ICE WATER.

ICE WATER

WAIT 2 MINUTES. TAKE COLD TEMP.

PINHOLE SUBMERGED

4. Record the temperature of the hot water, then *quickly invert* the tube into the cold water bath.

5. Wait 2 more minutes, then take the temperature of the cold water.

6. Use a 10 ml graduate to accurately measure the volume of air in the test tube at each temperature. Show your work.

21

ABSOLUTE ZERO (2) ◯ Kinetic Model (   )

1. Cut out the 2 Absolute Zero half-graphs along the dotted lines. Align the grids along the vertical axis, then tape together.

2. Accurately plot and circle 2 data points that summarize your results from the previous activity.

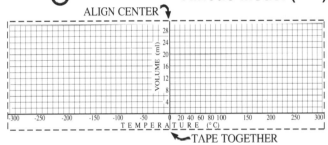

ALIGN CENTER

VOLUME (ml)

28 24 20 16 12 8 4

-300 -250 -200 -150 -100 -50 0 20 40 60 80 100 150 200 250 300
TEMPERATURE (°C)

TAPE TOGETHER

3. As air cools (under constant pressure), its volume shrinks in *direct proportion* to its drop in temperature. The average kinetic energy of its molecules decreases until they contract into liquid air with near-zero volume. This temperature where all thermal motion stops is called *absolute zero*.

   a. Extrapolate your two data points: extend a straight line back to zero volume and read the temperature at that point.

   b. Mark −273° C on your zero volume line. How close did you get to this accepted value for absolute zero? What limits your accuracy in this experiment?

4. Model absolute zero using a plastic bag and corn seeds as a kinetic model.

   a. Explain how you did this.

   b. Can air get colder than −273° C? Explain.

22

ABSOLUTE ZERO (3) Kinetic Model (   )

1. Set out a bowl and a jar nearly filled with warm water.

2. Clamp a clothespin handle just below the mouth of a test tube.

3. Drive off any moisture by heating it *gently* in an upright position. (CAUTION: Hot glass can burn you severely. Don't let its cool appearance fool you!)

4. Trap the hot dry air in the tube by inverting it at an angle. Continue heating for 1 minute, always keeping the mouth *below* the flame.

5. *Quickly* plunge the upended test tube by its handle into the bowl of water. Pour the jar of water over it until completely cool. Then trap *all* the water that enters with your thumb, and turn it right-side-up.

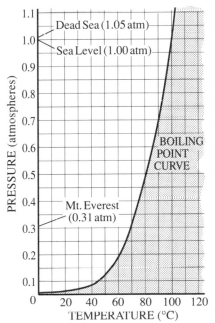

HOLD VERTI-CAL

6. Use a graduated cylinder, a thermometer and your graph to find how hot you heated the air in the tube. *(Hint: use Absolute Zero and one experimental point.)*

23

A COOL BOIL Kinetic Model (   )

1. In Activity 7 you boiled water.
   a. What boiling temperature did you record?
   b. According to this graph, what was your atmospheric pressure during that experiment?
   c. How would your results change at the top of Mt. Everest? On the shores of the Dead Sea?

2. Add 40° C water to a styrofoam cup.
   a. Is this water comfortable to touch? Can it boil at this temperature?
   b. Boil this luke-warm water inside a 3 cc syringe! Explain how you did it.

3. Water exists as a liquid *and* a gas only on the boiling point curve. Off this curve it is one *or* the other. Describe its phase(s) at:
   a. 40° C, 1 atm.   d. 90° C, 1 atm.
   b. 40° C, 0.08 atm.   e. 100° C, 1 atm.
   c. 40° C, 0.04 atm.   f. 110° C, 1 atm.

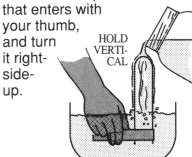

Dead Sea (1.05 atm)
Sea Level (1.00 atm)
BOILING POINT CURVE
Mt. Everest (0.31 atm)
PRESSURE (atmospheres)
TEMPERATURE (°C)

24

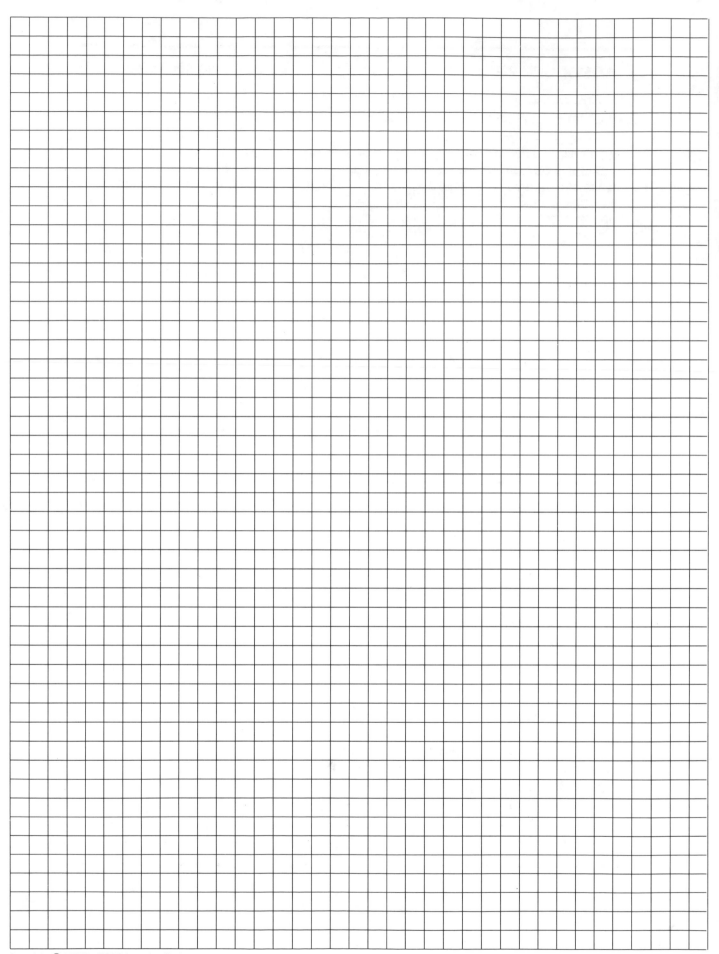

ACTIVITY 1

ACTIVITY 14
Relative Humidity Table

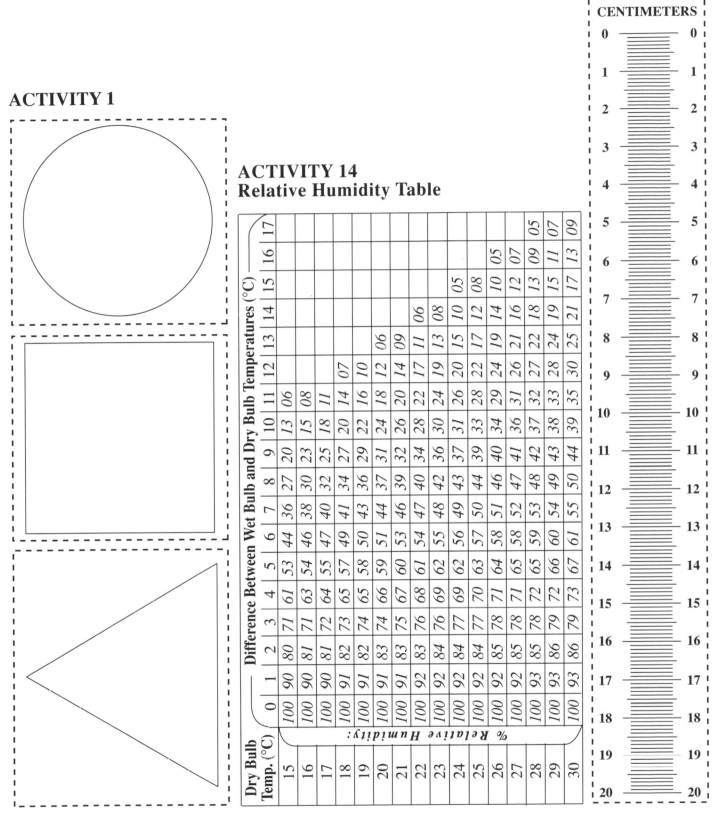

Difference Between Wet Bulb and Dry Bulb Temperatures (°C)

% Relative Humidity:

Dry Bulb Temp. (°C)	17	16	15	14	13	12	11	10	9	8	7	6	5	4	3	2	1	0
15							06	13	20	27	36	44	53	61	71	80	90	100
16							08	15	23	30	38	46	54	63	71	81	90	100
17							11	18	25	32	40	47	55	64	72	81	90	100
18						07	14	20	27	34	41	49	57	65	73	82	91	100
19						10	16	22	29	36	43	50	58	66	74	82	91	100
20					06	12	18	24	31	37	44	51	59	67	74	83	91	100
21					09	14	20	26	32	39	46	53	60	67	75	83	91	100
22				06	11	17	22	28	34	40	47	54	61	68	76	83	92	100
23				08	13	19	24	30	36	42	48	55	62	69	76	84	92	100
24			05	10	15	20	26	31	37	43	49	56	62	69	77	84	92	100
25			08	12	17	22	28	33	39	44	50	57	63	70	77	84	92	100
26		05	10	14	19	24	29	34	40	46	51	58	64	71	78	85	92	100
27		07	12	16	21	26	31	36	41	47	52	58	65	71	78	85	92	100
28	05	09	13	18	22	27	32	37	42	48	53	59	65	72	78	85	93	100
29	07	11	15	19	24	28	33	38	43	49	54	60	66	72	79	86	93	100
30	09	13	17	21	25	30	35	39	44	50	55	61	67	73	79	86	93	100

0 0
1 1
2 2
3 3
4 4
5 5
6 6
7 7
8 8
9 9
10 10
11 11
12 12
13 13
14 14
15 15
16 16
17 17
18 18
19 19
20 20

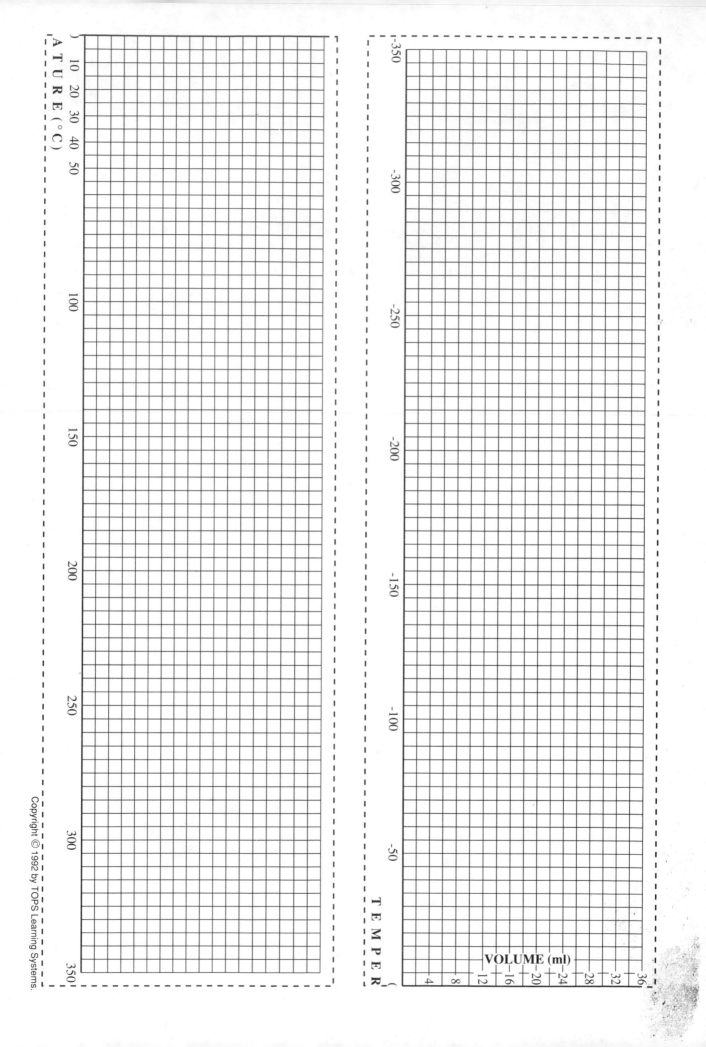

VOLUME (ml)

36 32 28 24 20 16 12 8 4

TEMPER

-350 -300 -250 -200 -150 -100 -50

A T U R E (° C)

10 20 30 40 50 100 150 200 250 300 350